REMAPPING AN ABLEIST WORLD

Disability and Oppression under Capitalism

Remapping an Ableist World examines the forces shaping our lives in an ableist capitalist world. It draws on examples including human enhancement and the organ trade to illustrate connections between able capitalist ways of life, impairment, disability, and oppression.

The book addresses ableness as a regime of power and oppression intrinsic to global capitalism and, as such, a system that touches all our lives, albeit in different ways. Vera Chouinard offers an intersectional analysis of the production of impairment and disability, drawing on autoethnographic and autobiographical methodologies, case studies of disability in the Global South and the Global North, and comparative accounts of processes such as the uneven development of disability law. Inviting readers to rethink the causes and consequences of the ableist capitalist order in which we find ourselves, *Remapping an Ableist World* reminds us that for our own well-being and that of generations to come we must forge a less destructive and more nurturing way of life.

VERA CHOUINARD is a professor emeritus of Earth, environment, and society at McMaster University.

Remapping an Ableist World

Disability and Oppression under Capitalism

VERA CHOUINARD

UNIVERSITY OF TORONTO PRESS
Toronto Buffalo London

© University of Toronto Press 2025
Toronto Buffalo London
utorontopress.com

ISBN 978-1-4875-0718-3 (cloth)   ISBN 978-1-4875-3574-2 (EPUB)
ISBN 978-1-4875-2487-6 (paper)   ISBN 978-1-4875-3573-5 (PDF)

Library and Archives Canada Cataloguing in Publication

Title: Remapping an ableist world : disability and oppression under capitalism /
    Vera Chouinard.
Names: Chouinard, Vera, author
Description: Includes bibliographical references and index.
Identifiers: Canadiana (print) 20240514335 | Canadiana (ebook) 20240514351 |
    ISBN 9781487524876 (paper) | ISBN 9781487507183 (cloth) |
    ISBN 9781487535742 (EPUB) | ISBN 9781487535735 (PDF)
Subjects: LCSH: Discrimination against people with disabilities. | LCSH: People with
    disabilities – Social conditions. | LCSH: Capitalism – Social aspects.
Classification: LCC HV1568 .C46 2025 | DDC 305.9/08 – dc23

Cover design: Sebastian Frye
Cover image: *Untitled* (1996), Judith Scott. Image courtesy of The Gallery of Everything.

We wish to acknowledge the land on which the University of Toronto Press
operates. This land is the traditional territory of the Wendat, the Anishnaabeg, the
Haudenosaunee, the Métis, and the Mississaugas of the Credit First Nation.

This book has been published with the help of a grant from the Federation for the
Humanities and Social Sciences, through the Awards to Scholarly Publications Program,
using funds provided by the Social Sciences and Humanities Research Council of
Canada.

University of Toronto Press acknowledges the financial support of the Government of
Canada, the Canada Council for the Arts, and the Ontario Arts Council, an agency of
the Government of Ontario, for its publishing activities.

Canada Council   Conseil des Arts
for the Arts   du Canada

ONTARIO ARTS COUNCIL
CONSEIL DES ARTS DE L'ONTARIO
an Ontario government agency
un organisme du gouvernement de l'Ontario

Funded by the   Financé par le
Government   gouvernement
of Canada   du Canada

Canada

To those who dare to dream of and forge a less oppressive world

Contents

Preface

We live in interesting but also quite frightening times. Developments such as a global pandemic and the rationing of health care in favour of younger, more able patients has reminded us of the fragility of human life and limits to our capacities to care for one another. This is also evident in harsh austerity measures that have slashed government income and other supports for the most vulnerable citizens in our societies – such as disabled people. These measures kill and in doing so help to reveal the "fault lines" in a global capitalist order that increasingly only "works" for the ultra-rich.

I have written this book because I am passionate about understanding impairment, disability, and ableness as outcomes of the contemporary capitalist way of life in which we are embedded. This is a challenging endeavour and one that this book just starts to address.

Although this book was written recently, my autoethnographic reflections draw on a lifetime of experiences – first as an able child and from the late 1980s onwards as a disabled woman and academic struggling against the odds for accommodations that would allow me to continue my work as a professor. The challenges I faced in this regard were a catalyst for conducting research into the lives of other disabled people. Recognizing my privileges as an academic woman located in the Global North, I was concerned to document and critique the more severe oppression with which so many disabled people are forced to contend.

Writing this book has been a contradictory experience. On the one hand, it has been very much a labour of love – an opportunity to address one of the key issues of our times – namely, why disabled people remain so oppressed and excluded in able capitalist societies and what we can do about it. Relatedly, it creates possibilities for "speaking truth to power" in empowering ways. For example, by acknowledging that austerity measures kill and/or further disable those with impairments and/or illnesses, it becomes possible to name such measures for what they are: a cleansing of contemporary societies of some of its most vulnerable members. On the other hand, sharing personal accounts of impairment, disability, and ableness and exploring or remapping how these are shaped by forces within global capitalist societies can be deeply distressing. As I know from my own autoethnographic work, it forces you to relive the trauma of being oppressed and devalued. In other words, the scars of disability and ableness run deep. There is also the possibility of losing hope that linkages between impairment, disability and ableness and capitalism mean that we are unlikely to change disabled people's lives for the better.

I encourage you to read this book in a hopeful and passionate spirit. For while much remains to be done in building more inclusive and enabling societies, we are also learning to think more expansively about why a global able capitalist way of life is failing those who embody differences such as impairment and disability. We also need to learn more about how ableness harms not only disabled people but also those with able bodies and minds.

Acknowledgments

Many people helped to make this book possible. First, I would like to thank the disabled and chronically ill people who generously shared their life stories with me, my research assistants, and my graduate students – we learned so much from you about ableness and disability. I also would like to thank Dr. Linda Peake and Karen DeSouza for helping to facilitate research conducted on disability in Guyana. This work would not have been possible without your support and guidance. Discussions with students in my graduate and undergraduate classes helped to fine-tune ideas about the dynamics of an able, neo-liberal capitalist world, including its oppressive consequences.

I am indebted, professionally and personally, to those who assisted me during my very lengthy and difficult struggle for accommodation as a disabled professor. These included lawyers in private practice and in organizations such as CAUT (Canadian Association of University Teachers) and equity officers at my university. Among other things, they were able to reassure me that forcing me to quit or to work only on terms that discriminated against me on the grounds of disability were violations of my human rights under Canadian law.

I was fortunate to find other allies on and off campus who helped to bolster my determination and hopes to help build a more inclusive world. These allies included faculty and staff (approximately thirty people including myself) who were members of a committee that was aiming to draft an accommodation policy for the university. The policy was drafted by the mid-1990s but Senate approval was delayed until the end of the decade – reportedly because administrators left the report sitting on their desks. Nonetheless, the fact that this struggle was underway offered hope that we could make our own and other universities more inclusive of diverse people. Being part of an intellectually and politically vibrant women's studies community also provided invaluable support. Not only did these colleagues provide support and encouragement when I was dealing with difficulties associated with being a diverse woman within the academy, but they also delivered thought-provoking talks. Allies off campus included disability activists advocating for passage of an Ontarians with Disabilities Act. Working with local branches of the committee and taking part in events such as protests and rallying for the act at Queen's Park (the seat of provincial government) helped to underscore the pressing need for such measures.

Jodi Lewchuk, acquisitions editor at University of Toronto Press, expertly steered the book through the design, editing, and production phases. Beth McAuley, freelance editor, and Janice Evans, associate managing editor at University of Toronto Press, helped to ensure that the meaning of the text was clear, that it was free of grammatical and spelling errors, and that the references were correct and complete. I want to extend a special thank you to Jodi, Beth, and Janice for assisting with some of this work when I was too ill to complete it.

Finally, my love and heartfelt thanks to my family. To my partner, Ed – you have never wavered in your love, support, and care no matter how difficult our lives have been. To my wonderful daughter, Renee, and my son, Alexander, thank you for believing in me and for all your love and support. I am so very proud of the awesome people you are!

To our new granddaughter, Esme, to her mother, Lyndall, and to her sister, Dilana – thanks for all the joy you bring into our lives.

Thanks as well to our furry family members, Dante, Chloe, and Maybelle, for all their love and support. You mean the world to us!

REMAPPING AN ABLEIST WORLD

1 Introduction: Remapping an Ableist World

We are living through tumultuous times. Our world seems to career from one existential crisis to another. There is, for example, the climate crisis that is killing life as we know it on earth. This is rooted in a capitalist world order that for too long has prized profit-making from the exploitation of oil and gas resources – the burning of which is known to diminish human and planetary health – over the development of green energy alternatives. Governments are complicit in supporting continued reliance on oil and gas. A March 2023 report in *The Guardian* (Horton, 2023), for example, revealed that governments in the UK have transferred 20 billion more pounds to the oil and gas sector than to renewable energy since 2015. At least one-fifth of that money went to new oil and gas extraction projects. We are also enmeshed in a culture of consumption that still too often generates toxic wastes such as the plastic particles that have made their way into the bodies of humans and other species such as marine life.

We are living through a global pandemic and experts predict more are on the way. As of 5 April 2023, the WHO (World Health Organization) reported 762,201,169 confirmed cases of COVID-19 and 6,889,743 deaths worldwide. People also are becoming impaired and unable to work because of so-called "long COVID." In March of 2022, a US Government Accountability Office (2022) report estimated that long COVID had affected as many as 23 million Americans of whom some one million had been pushed out of work. The pandemic has also pushed our health care systems beyond their limits, as evidenced, for example, by the rationing of care to the more able-bodied in countries as diverse as the United States and Democratic Republic of the Congo.

To these crises we must add spiralling costs of living amid the growing polarization of wealth, the rise of more precarious working conditions, and a general deficit in the reproduction of labour power. As

Fraser (2022) pointed out in her recent book *Cannibal Capitalism*, we are witnessing a system that is devouring its own means of existence by, in part, failing to provide the reproductive care that is a necessary condition of a capitalist way of life based on the exploitation of workers. Such developments have important implications, including a decline in standards of living and life expectancy in most nations of the world. In 2022 the United Nations reported that its Human Development Index, which measures standards of living, health, and education in a country, had declined in 90 per cent of countries for the second year in a row for the first time in thirty-two years (United Nations Development Programme, 2022). In short, our world is becoming a harsher and more difficult place to survive in, and this is taking its toll as evident, for example, in the mental distress that studies suggest is related to conditions of life such as precarious work.

There is, then, an urgency to furthering our understanding of the dynamics of the able global capitalist order that we find ourselves in. This book seeks to do so by exploring what it is like to live in today's ableist world. Here, I draw on personal accounts, including my own, to help shed light on how ableness and embodied differences such as "being mad" shape our ways of being in the world. This book is also concerned with the uneven production of impairment and disability in our contemporary neo-liberal capitalist order. The fact that 80 per cent of the world's disabled people live in poorer countries of the Global South is a sobering reminder of just how geographically uneven and unjust the production of impairment and disability is.

In this book, I aim to unsettle notions that ableism, as a regime of power and privilege, pertains only to disabled people's lives. I argue, rather, that we are all creatures of ableness, although this can be experienced in significantly different ways depending, for instance, on where in the world we live and whether we are able. We are also creatures of an increasingly harsh global capitalist world that has "ramped up" the exploitation of our capacities to work hard to the point where those very capacities are placed at risk. At the same time, we are witnessing the rise of right-wing populist governments, which are putting in place austerity measures that harm and kill vulnerable groups such as the elderly and disabled. We are thus experiencing what I will later term a "violent cleansing," although few people call it out as such. It is sobering that this is happening in both the Global North and Global South and that those who benefit from austerity policies and associated tax breaks are the ultra-rich.

This book aims, then, to understand ableness or ableism as central to all of our lives and, in doing so, to help us situate the production of impairment and disability as arising from a neo-liberal global capitalist world that often seems "hell-bent" on destroying life as we know it. One of the most frightening manifestations of this is climate change and, while this is not the primary focus of this book, I do recognize that this is part of contemporary capitalism's contradictory tendencies that are running amok and causing impairment, disability, and death.

In writing this book, I also hope to contribute to the collective project of remapping an ableist world. By this I mean rethinking the dynamics of an able neo-liberal capitalist world and reimagining possibilities for social change. I do this, in part, through autoethnographic accounts of growing up in an able and misogynistic world and life as a disabled female professor and woman. But I also do this through a consideration of the ways in which ableness, impairment, and disability at national and international scales are matters of social injustice. This includes difficulties in advancing disabled people's human rights and well-being through disability human rights law. We need to think, I argue, beyond the limits of law if we wish to advance disabled peoples' rights and well-being in our contemporary capitalist world.

As a woman who has been visibly disabled for most of my adult life, writing this book has been both a labour of love and rooted in a sense of outrage at how we are treating vulnerable people in our contemporary capitalist world. This includes growing numbers of workers employed in precarious jobs and those experiencing exclusion from the labour market and spaces of everyday life such as many disabled people. We owe it to ourselves and to future generations to help create a world that supports human and other beings in all their diversity. And we need to find ways of sharing wealth rather than, as is the case today, hoarding wealth for the benefit of a wealthy elite that grows progressively smaller.

In the remainder of this chapter, I map out how ableness, impairment and disability will be addressed in the chapters that follow. But first key terms are defined.

Definitions

Throughout this book I use the terms "ableness" and "ableist/ableism" interchangeably to denote a regime of power, privilege, and oppression pervasive in our contemporary capitalist order. It is a regime that values and privileges able bodies and minds and devalues impaired ones.

Ableness finds expression in a range of practices aimed at enhancing human abilities.

Impairment refers to functional differences in the human body and mind. But impairment is also culturally constructed, for example when a visible impairment such as missing limbs is interpreted as "tragic."

Disability refers to the processes whereby those with impaired and ill bodies and minds experience their embodied differences (e.g., through pain and mobility differences) and encounter a range of barriers to their inclusion, well-being, and participation in spaces of everyday life. These barriers include physical ones such as buildings that are only accessible via stairs, and social and attitudinal barriers such as exclusion from paid employment and devaluing disabled people's abilities and lives, respectively.

Neo-liberalism has been defined by Harvey (2018) as a political project to advance the wealth and power of capitalists at the expense of members of the working class. Originating in the 1960s and 1970s, the term involved diminishing government intervention in the capitalist economy (sometimes referred to as a return to an "unfettered free market"), a dismantling of the post-war welfare state in favour of austerity measures (e.g., reducing government expenditures on income assistance and social services), and, through globalization, shifting production to less affluent nations where profits and capital accumulation could be bolstered by cheaper labour costs. This, along with a decline in manufacturing in countries such as the United States, helped to discipline workers who were seeing their once powerful unions grow weaker – undercutting their bargaining power.

I also want to comment on why I use "disabled persons" rather than the people first language favoured in North America (i.e., people with disabilities). This is because "disabled persons" helps to foreground the notion that people with impairments and illnesses are disabled by society and face barriers to inclusion in a range of different places.

Remapping an Ableist World

A central project of this book is to challenge or "remap" our understanding of and responses to ableness, disability, and impairment. This is done both literally and metaphorically. Literally, I call for greater attention to the uneven production of impairment itself, and for ongoing efforts to put the places in the Global South where most disabled people live on agendas for research and action on ableness, disability, and impairment.

Remapping is also used in a metaphorical sense to denote, for example, expansive conceptions of ableness, disability, and impairment as outcomes of the dynamics of global capitalist societies. In a related vein, this remapping includes expanding the processes and phenomena seen to be associated with ableist regimes of power and privilege. Examples in this book include considering the pursuit of human enhancement, the illegal organ trade, and the limits to law as a terrain of struggle to empower disabled people, particularly in the Global South.

My autoethnographic reflections in this book are also a remapping or reworking of my experiences of ableness, disability, and impairment in different places of everyday life (see chapter 3 on multiple layers of oppression in my academic workplace and chapter 4 on my own and others' experiences of negotiating people and places with bipolar disorder).

I liken my experiences of struggling for accommodation as a disabled female professor to experiences of being in a baffling and illogical world akin to that of Alice in Wonderland. Ideally, autoethnography does not recount the personal and intimate details of life for their own sake but for what they can reveal about the political, economic, and cultural forces shaping our lives. Thus, I hope that readers will see this book as an invitation to engage in the autoethnographic musings that help to make the personal political.

Synopsis of Chapters

Chapter 2 underscores the importance of thinking about ableness, impairment, and disability as matters of social injustice. This has several advantages. One of the advantages of this is that it allows us to think more relationally about the processes through which people with impairments and illnesses become disabled (E. Hall & Wilton, 2017). Another advantage is that it avoids constructing a dichotomy between impairment as a physical state and disability as a social state as is done in traditional social models of disability. A related advantage is that it allows us to reflect upon the geographically uneven production of ableness, impairment, and disability and why it is that most disabled people live in poorer countries of the Global South. This, in turn, raises issues of social and spatial injustice associated with neo-colonial relations of power and oppression. It also emphasizes the importance of context in shaping the lives and struggles of disabled people. Rather than treating disability as a universal condition as in Western disability discourses of individual rights, this approach stresses that context and place are of central importance in understanding experiences of ableness, impairment, and disability. In short, I argue that framing ableness, impairment, and disability as matters of social injustice helps us to think more expansively about the processes at work in perpetuating an oppressive able neo-liberal capitalist world order.

The autoethnographic writings about my experiences of being a disabled female professor have been an important part of my journey through life with physical and mental impairments/illnesses. As I note in more detail in chapter 3, autoethnography is about sharing aspects of one's life that shed light on the social and cultural dynamics of the world in which we live. It is not sharing the personal for its own sake, but sharing because of what we can learn about the dynamics of our societies. It thus seems fitting to include aspects of my own experiences in chapters 3, 4, 6, and 7.

Chapter 3 provides a more expansive account of living in an ableist world than in my previous autoethnographic writings. Here, I provide an account of what it was like to grow up in an able and misogynistic environment, and then I go on to consider what can be learned from my experiences as a disabled female professor and through reflection on my experiences from the vantage point of a senior disability studies scholar. Among other things, this chapter helps to illustrate how ableness is at work in both able-bodied and disabled people's lives. In chapter 4 I include an autoethnographic account of experiences of bipolar "madness" and then go on to consider what we can learn as well from autobiographical accounts of life with this illness. I argue that experiences of bipolar disorder are not straightforwardly "irrational" but, rather, involve a careful reading of people and places as dangerous or safe. In this context, valid details about the changes to the environment and people one is interacting with in it may be discerned, but the interpretation of the meaning of those attributes may be, in part, delusional. One basic point to take away from this is that dichotomizing sanity and madness does not do justice to the experiences that people with bipolar share. This ignores the fact that those experiences include, for example, moments of keen insight and logic.

Chapters 5, 6, and 7 shift the analytic and empirical focus on ableness, impairment, and disability from the personal and interpersonal scales to the national and international. This is not intended to discourage readers from reflecting on how they are personally caught up in forces unleashed by global able capitalism but, rather, to begin to provide the conceptual "scaffolding" we need to better understand how the personal and interpersonal are always simultaneously situated in national and global contexts. In chapter 5, I consider some of the ways in which our contemporary global able neo-liberal capitalist order perpetuates oppression, ableness, impairment, and disability. I argue that this takes a variety of forms, including the hyper-exploitation of our capacities to labour, the rise in precarious jobs, and austerity measures that harm and indeed kill vulnerable people such as elderly and disabled persons. I argue that this ought to be understood as a kind of "violent cleansing" on the part of right-wing populist governments – in which more and more people are treated as unimportant and "disposable."

Chapter 6 goes on to consider the commodification of human embodiment itself, from the consumption of goods and services that promise ableness and health through to modification of the body itself (inside and out). This too raises matters of class and social injustice in terms of who can and who cannot afford to enhance their abilities. I illustrate this regarding one of the "darker" sides of human enhancement practices, namely, the illegal trafficking and transplantation of human organs. It also raises questions about how far efforts to enhance our abilities should go. Should we attempt to eliminate impairment and disability or is there a case to be made for including diverse forms of embodiment and abilities in society and space?

In chapter 7, I explore the potential and limits of disability human rights law, as developed in the Global North, for advancing the inclusion and well-being of disabled people in countries of the North and South. I argue that this is based on both an individualistic and universal conception of the lives of disabled people. It is individualistic in the sense of assuming that disabled individuals are bearers of uniform rights and universal in the sense of positing a common experience of ableness, impairment, and disability.

There are also practical reasons why the existing international and national legal systems fail to deliver justice and inclusion. One is the legacy of neo-colonial relations of power which disadvantage and exploit poorer nations of the Global South in the struggle against ableness, impairment, and disability. Another is the underfunding of the transnational justice system in regions such as the Caribbean (Chouinard, 2018). There is also the disjuncture between international and national legal systems. The former, particularly the UN Convention on the Rights of Persons with Disabilities has been ratified by many nations, but especially in the case of poorer nations, it is difficult or impossible to implement in practice. Using a case study of legal struggles in Guyana, I argue that we need to "think outside the box" of the current legal system and law itself to effectively challenge the marginalization, oppression, and disabling of people with impairments and illnesses worldwide.

Chapter 8, the concluding chapter of this book, discusses the implications that the arguments made in this book have for our understanding of the production of ableness, impairment, and disability and for efforts to build a less able but paradoxically more enabling global order. It also considers the political and policy implications of these arguments. I argue that our approaches need to be formulated as being about more than individual rights and include issues of distributional injustice between the Global North and Global South. Finally, I briefly discuss key challenges in remapping the ableist world we share.

2 From Disability to Social Injustice: Reframing Ableness, Impairment, and Disability Issues

The last few decades have seen important advances in how scholars and activists conceptualize ableness, impairment, and disability. This chapter focuses primarily on recent contributions to the interdisciplinary field of disability studies. Drawing on the work of selected feminist, political-economic, materialist, and queer scholars, I argue for a conceptual framework that allows us to reframe questions about ableism, impairment, and disability issues as matters of social injustice at geographic scales ranging from the personal to the global. Among the advantages of this approach are that it allows us to think more relationally and intersectionally about disability and to move away from neo-liberal capitalist discourses that promise, but seldom in practice deliver, individual equality through human rights measures. This is not to argue that struggles to advance disabled people's individual rights are futile, they have, for example, considerable symbolic and political meaning, but it is to argue that we need to think more expansively about the causes and consequences of impairment, disability, and ableness in our neo-liberal capitalist world. By framing the uneven sociospatial production of disability and impairment as part and parcel of an able neo-liberal capitalist regime, which is fundamentally materially unjust, we can begin to imagine and work towards an alternative, more empowering social order.

From the Medical to the Social Model and Beyond

The late twentieth century saw a shift from the medical model of disability to the social model of disability. This had its roots in the emerging disability movement as well as in disability studies scholarship. The importance of this shift lay in no longer attributing disability to a body in need of a "cure" through biomedicine to a recognition that disability was societal in nature and reflected attitudinal and environmental barriers to the inclusion of disabled people. This shift has been discussed elsewhere (see, e.g., Chouinard et al., 2010; Oliver, 1990; Park et al., 1998; Shakespeare, 2006, 2011), so I will not further revisit it here.

The twenty-first century has seen important advances beyond this shift. These include more embodied but also still social conceptions of impairment and disability (e.g., Erevelles, 2011), growing interest in ableism as a regime of power and privilege (e.g., Campbell, 2009; Goodley, 2011), feminist materialist efforts to reconceptualize the processes involved in the marginalization of disabled people (Garland-Thomson, 2011), and queer theorists' reflections on how intolerance towards disabled embodied subjects is related to processes of devaluing anomalous bodies more generally (McRuer, 2018) and efforts to think about disabled people's experiences more intersectionally (e.g., Valentine, 2007).

Interest in forces shaping disabled people's lives and struggles in the Global South has also been on the rise in critical disability studies as seen, for example, in the collection edited by Grech and Soldatic (2016).

In this chapter I argue for a critical, materialist, and feminist approach to understanding forces promoting what McRuer (2018) has dubbed "compulsory able-bodiedness." I also consider the geographically uneven production of impairment and disability associated with contemporary "ableist" neo-liberal capitalist societies. I begin, in the next section, by discussing the senses in which my conceptual approach is "materialist."

Thinking Materially About an Ableist World

The conceptualization developed in this chapter is "materialist" in at least three ways. First, as in Gleeson's (1999) now classic book *Geographies of Disability*, it aims to situate ableness, impairment, and disability in the materialist context of capitalist society. Unlike Gleeson's historical materialist approach, however, this book is not primarily concerned with how transitions between modes of production helped to transform disabled people's places in society and space. The emphasis, instead, is on contemporary global capitalism and some of the dangerous forces it has unleashed – forces that are especially threatening to disabled people's lives but impact able-bodied people as well.

This conceptualization is also materialist in the feminist sense that experiences of ableness, impairment, and disability are understood as embodied, intersectional, and relational. This allows us to consider bodily/mental differences such as chronic pain and "madness" as part of experiencing an ableist world. The need for feminists to develop more intersectional approaches to theory and analysis was first flagged by Crenshaw (1989), a Black feminist legal scholar. She argued that rulings on cases of discrimination against Black women in the US justice system were failing to acknowledge that these could involve discrimination based on multiple identities such as being a woman of colour. She called on scholars and activists to shift away from what she termed "one-axis" approaches to discrimination (such as when US court rulings recognized only that women could be oppressed as women or as Black persons but not both).

The last three decades have seen growing interest in intersectionality as a theory and method that can enhance our understanding of why and how multiple embodied differences become bases of discrimination, oppression, privilege, and power. A special 2013 issue of *Signs: Journal of Women in Culture and Society*, for example, explored intersectionality as an interdisciplinary field of study. The introductory article, written by Cho et al. (2013), discussed three main lines of enquiry in what the authors termed this "burgeoning" field of study. The first line of enquiry they described in the following way:

> The first approach applies an intersectional frame of analysis to a wide range of research and teaching projects. Aggregated together in this category are undertakings that build on or adapt intersectionality to attend to a variety of context-specific inquiries, including, for example, analyzing the multiple ways that race and gender interact with class in the labor market; interrogating the ways that states constitute regulatory regimes of identity, reproduction, and family formation; developing doctrinal alternatives to bend antidiscrimination law to accommodate claims of compound discrimination; and revealing the processes by which grassroots organizations shape advocacy strategies into concrete agendas that transcend traditional single axis horizons. (p. 785)

This line of enquiry can be summarized as material, expansive, contextual, and activist – the goal is to expand the embodied differences considered, the contexts in which multiple ways of interacting are examined, and to explore whether and if so how grassroots organizations are developing advocacy strategies that address compound discrimination and oppression.

Cho et al.'s (2013) second line of enquiry focused on discursive debates about intersectionality as theory and method. Here, the concern is to address fundamental questions such as whether there is an essential subject of intersectionality and whether or not that subject is static in time and space or dynamic and fluid. This line of enquiry can be described as focusing on the theoretical and methodological assumptions that inform intersectionality studies.

In the final line of enquiry, Cho et al. (2013) emphasized praxis in keeping with the view that the point of critical studies is not just to explain why the world is as it is but to use that knowledge to create a better and more empowering future. In this context there is a primary concern to think through what the implications of intersectional research are for concrete studies of social change and what some of the associated "best practices" for community organizing might be.

In this book, I am primarily concerned with the first and arguably most expansive line of intersectional enquiry and, to some extent, the third line of enquiry. In chapter 3 I provide an autoethnographic account of how multiple embodied differences (such as age, gender) and others' reactions to bodies and minds that differed from their own created spaces of childhood that were ableist and misogynistic, even if largely invisibly so through a child's gaze. This chapter also shares some of my experiences as a disabled female professor in the academy and as a disabled woman beyond its walls. Here, I illustrate some of the ways in which the academy and wider communities are profoundly ableist and gendered and how the inscription of other devalued identities (such as, in my case, as being "too radical" and "not doing real research") further complicated the forms of oppression and exclusion that I encountered. Chapter 4 continues to explore the embodied, intersectional materiality of ableism, impairment, and disability at the intra- and interpersonal scales but now with reference to experiences of bipolar disorder (also known as manic depression). Drawing on my own experiences as well as on autobiographies by others with this illness, I show how we can help develop a more nuanced understanding of what "madness" entails and challenge stereotypical understandings of "mad" ways of being in the world as straightforwardly "irrational" and/or dangerous responses to people, places, objects, and events.

Thinking intersectionally about ableism, impairment, and disability is important throughout this book. The final chapters do so, however, primarily at the societal and global scales in order to make connections between our contemporary global capitalist order and its geographically uneven development, the production of impairment and disability, and the pervasiveness of ableist privilege and power.

Remapping an Able World and the Production of Impairment and Disability as Issues of Social Justice

Disability activists and scholars have a long history of trying to reframe impairment and disability as issues of social justice (e.g., Gleeson, 1999; Oliver, 1990; Titchkosky, 2008). Gleeson (1999), for example, used Young's (1990) theory of the multiple faces of oppression (e.g., exploitation and exclusion) to help understand disabled peoples' changing places in society and space. Dodd (2016), drawing on Fraser's (2022) work, argued that disability studies could be enriched by using concepts such as cultural misrecognition and marketization of services to understand the oppression and injustices that disabled people face. He argued that Fraser's bivalent approach to understanding processes of oppression as cultural and economic is a promising basis for charting directions forward in disability studies. A key advantage of such an approach is that it allows exploration of both cultural and political-economic forces shaping disabled people's lives. In doing so, it helps to reframe ableness and disability as matters of material and cultural injustices that are at the heart of our able neo-liberal capitalist order.

Efforts to address such aspects of oppression, in scholarly and political practice, have, however, tended to be couched in terms of advocating for equal human rights for disabled people – to employment, access to buildings and services, and participation in social life on terms comparable to those of non-disabled people. This reflects the dominance of Global North discourses and practices in constructing disabled people as individual bearers of legal rights. More recently, however, disability scholars such as Grech and Soldatic (2016) have called for more expansive perspectives on the processes contributing to the marginalization of disabled people. It has been noted, for example, that experiences of disability in countries of the Global South have been relatively neglected and that phenomena such as the uneven production of impairment, concentrated in the Global South, need to be incorporated into accounts of disability. Mitchell (2015) and the International Labour Organization (ILO; n.d.-a), for example, pointed to how unsafe working conditions in countries of the Global South contribute to death and impairment as in, for instance, the Rana Plaza textile factory collapse in Bangladesh in 2013. This collapse killed over 1,100 people and injured or impaired over 2,000 others. Factors contributing to the collapse included unsafe building practices and standards, and management practices that forced labourers to continue to work despite their fears that the building was unsafe. Accidents, including textile factory fires, have also claimed people's lives or abilities (ILO, n.d.-a; Mitchell, 2015). Such events can be seen as outcomes of the injustices inflicted on poor labouring bodies and minds – denying them any control over the labour process even if this means putting their lives and capacities to work at risk.

In the medical model of disability, as discussed above, impairment and disability are regarded as individual mind and/or bodily states and the goal is to use biomedicine to mitigate symptoms or provide a cure. This is an individualist model and as such does not encompass questions of social justice. In contrast, the social model of disability, also discussed above, identifies attitudinal, social, and physical barriers to inclusion as a social problem that, at least implicitly, requires collective action to create more enabling environments. The social model in this way gestures towards the possibility of disability being a social justice issue. Having said this, however, there remains a need to further explore the links between ableness, impairment, and disability and social injustice.

While the social model directs our attention to barriers to the inclusion of disabled people that exist, it does not, as Oliver (1996) insisted, offer a theoretical framework that can help us understand why such barriers to disabled people's inclusion in society and space arise and persist. In other words, this doesn't help us explain the processes involved in creating societies and spaces of life that exclude disabled persons. Not surprisingly, then, disability studies scholars, such as Oliver (1990) and Gleeson (1999), turned to other theoretical/methodological approaches to explain their findings. In Oliver's (1990) case, it was a Marxist account of how capitalist societies devalue disabled people, and in Gleeson's (1999) case, it was a historical materialist approach that linked the increasing marginalization of disabled people to transitions between feudal and capitalist modes of production (such as the rise of a factory system of production and labour practices that characterized disabled people as inferior, slow workers).

The social model, with its dichotomous understanding of impairment as a physical state and disability as a social state, also jettisoned opportunities to explore the production of impairment itself as a matter of social injustice and embodied experiences of impairment such as pain and fatigue as part of the experience of disability in society and space. Furthermore, with its focus on barriers as a matter of violating human rights, the model did little to encourage critical accounts of ableness as a regime of privilege and power. In the next section, I consider more recent scholarship that has attempted to address these facets of disabled people's lives.

Feminist, Crip/Queer, Political Economic, and Neo-colonial Perspectives on Impairment, Disability, and Ableness as Matters of Social Injustice

More recent work by scholars using feminist, "crip"/queer, political-economic, and neo-colonial frameworks arguably offer more expansive conceptions of impairment, disability, and ableness (or ableism) as matters of social justice and injustice. Mladenov (2015), for example, built on Fraser's analysis of the affinity between the rise of second-wave feminism and neo-liberalism. He wrote:

> A number of disability scholars have highlighted the link between the rhetoric, the logic, the principles and aims of the DPM [Disabled People's Movement] and the neoliberal emphasis on consumerism privatisation, deregulation and decentralisation ... Looking at a similar convergence between second-wave feminism and neoliberalism, Nancy Fraser (2013, 218) poses the question of the link between the two thus: "Was it mere co-incidence that second-wave feminism and neoliberalism prospered in tandem? Or was there some perverse, subterranean elective affinity between them?" Fraser's (2013, 244) answer is that the underlying affinity between feminism and neoliberalism was the critique of traditional authority. Her insight could be applied to the DPM as well, because the critique of traditional authority has been at the core of the DPM's struggles. Indeed, these struggles have generally sought to promote social justice rather than marketisation, and citizenship rather than consumerism. (p. 450)

McRuer (2018), a crip/queer scholar, has argued that compulsory able-bodiedness and austerity are intrinsic features of contemporary neo-liberal capitalist societies. It follows that dis/ability is always the invisible "elephant in the room" when it comes to making sense of the dynamics of our societies and the issues of social injustice that we face. Quoting the work of Puar (2009, 2010), Goodley (2014) developed a related argument about why what some term "anomalous bodies" (e.g., disabled, queer, raced) are exceptionalized while "trans-humanist ones" (e.g., that use special prosthetic legs for sports) are highly valued. He stated:

> Wherever we lie in relation to the dis/ability binary we are all found lacking: there is debility amongst us all. I agree with Puar that "all bodies are being evaluated in relation to their success or failure in terms of health, wealth, progressive productivity, upward mobility, enhanced capacity." While some bodies and populations are deemed more precarious than others, we are all debilitated in and by neoliberal capitalism. (Goodley, 2014, chap. 6)

Or, in other words, just like none of us can live "outside" capitalism nor should we imagine – even if we are judged "able" in body and mind – that we can escape being disciplined by compulsory able-bodiedness. I illustrate this, in part, in chapter 3 where I recount some of my experiences as an able female child.

What are the implications of this line of argument for reframing impairment, disability, and ableism as issues of social injustice? First, and perhaps most obviously, that all of us are harmed, albeit to varying degrees, by the ableness demanded in neo-liberal capitalist societies and spaces of life. This is injustice in the sense of intolerance towards diversity in bodies and minds and in capacities. Second, that there are distributional issues of justice that need to be considered. What can be done, for example, to address the global uneven development of capitalist societies that gives rise to a concentration of ill, impaired, and disabled people in poorer countries of the Global South (they account for 80 per cent of the more than one billion disabled people worldwide)? Why are disabled people worldwide most likely to be consigned to lives of poverty and how can this be challenged? Then, there are representational issues of social injustice. For instance, why is it so hard for us in contemporary able neo-liberal capitalist societies to accept the "normality of doing things differently" (Hansen & Philo, 2007), such as wheeling instead of walking?

As McRuer (2018) has argued, aspiration and austerity are also central to the workings of an able neo-liberal capitalist social order. He wrote:

> Aspiration has basically been codified … in the era of neo-liberal capitalism, as an individualist, libertarian concept oriented around personal achievement and merit. This sense of aspiration is in many ways now compulsory; it secures consent to the dominant economic and political order, effectively dividing people from one another and short-circuiting any kind of class analysis. (p. 176)

Deep and harsh government spending cuts, also known as austerity measures, have been central in disciplining citizens; especially those least able to conform to able-bodied and aspirational ideals. But they have done more than discipline citizens: they have also extinguished hope for a better future and, in some cases, extinguished life itself. McRuer (2018) wrote on disability in Liz Crow's (UK artist and activist) ambitious art installation *Figures* and noted that many of the 650 figures she created out of clay and mud from the Thames River and displayed for public viewing were associated with the stories of disabled people in the UK. Some of those stories were truly heartbreaking: People who committed suicide rather than having to continue being stressed and disheartened by trying to appeal cuts to their benefits. A kind, elderly man who died of starvation because of too little income to purchase food. Another HIV-positive man who died when his life-saving medication ran out. And the human harm caused by austerity includes discrimination against disabled jobseekers. Ryan (2018) commented on this in the UK context:

> Meanwhile, the benefit sanction regime has been found to be discriminating against disabled people: disabled jobseekers are 53% more likely to have their money docked than a claimant who isn't disabled. And only last month the National Audit Council [NAC] found that the government has falsely underpaid about 70,000 disabled people on ESA, with some now owed as much as £20,000. The NAC found that the government knew about it in 2013 but did nothing. (para. 11)

She continued:

> This is Britain's rigged safety net, where disabled people must jump through hoops to get scraps of support, but the government can remove it anytime it wants.
>
> A few days later, Michael [a disabled informant] emails to tell me his mum's had a stroke – another worry on top of the rest. The bills are mounting, and it's getting too much to deal with. Michael wonders what the point of ministers making these cuts was, and for the life of me, I can't find one. "There's no public gain," he says. "Just a great deal of loss to myself and others like me." (paras. 12–13)

It is important to acknowledge that such measures are widespread in our times and places. A report on planned social spending cuts in the United States under the Trump regime noted that rather than helping low- and middle-class families through policies such as childcare and raising the minimum wage, the administration and its Republican base were pushing for severe cuts to benefit programs. The Trump administration also worked to limit access to those programs through making the running of the programs more difficult and imposing tougher work requirements. This is a form of austerity that only benefits the rich, with savings from these cuts (amounting to US$1.5 trillion) being reserved to pay for tax cuts to wealthy individuals and corporations (Hamm et al., 2018). In contrast, the 2021 government spending plan under the Biden administration included $400 billion for free and universal preschool for all three- and four-year-olds and $150 billion for building one million affordable housing units. However, key social spending initiatives were lost as the plan moved through Congress. These losses included paid family leave, which was removed entirely. This was a setback for progressives who had hoped the US would follow the lead of most other nations in providing paid time off for new parents. The

US is one of only eight countries without national paid maternity leave. A plan to reduce prescription drug prices was also slashed – something that outraged the American Association of Retired Persons, the largest organization focused on elderly Americans (BBC, 2021).

The provincial Conservative government of Doug Ford, in Ontario, Canada, has cut government spending on a wide variety of programs. In terms of social services alone, the following cuts were made during its first year in office:

- Cut $1 billion from social services across the board
- Scrapped Basic Income Pilot Project
- Cancelled $1 increase [in the] minimum wage
- Cut Workplace Safety Insurance Board payments to injured workers by 30 per cent
- Killed Bill C-148, which provided part-time workers the same pay as full-time workers, guaranteed 10 days off (2 days paid) and more
- Removed rent control for new units
- Severed library services funding in half
- Ended the Roundtable on Violence Against Women
- Slashed $84.5 million funding for children and at-risk youth, including children's aid societies
- Cut $15 million from the Ontario Trillium Foundation. (Syed, 2019, Social Services, para. 1)

Other cuts have been made to clean energy projects, medical care, education, the arts and music, and university tuition fees. The Ford government has also slashed funding to Legal Aid Ontario and legal clinics that serve persons who have low or even no income, including disabled people. The executive director of the ARCH Disability Law Centre, a legal clinic specializing in disability issues, summarized these cuts in the following way:

> Among these austerity measures was a significant reduction to Legal Aid Ontario's budget in the amount of $133 million during this current fiscal year, with a larger funding decrease next year. Provincial funding for immigration and refugee legal services was explicitly cut. In response to this budget reduction, Legal Aid Ontario undertook a number of measures which, in addition to significant reductions in immigration and refugee services, included changes and reductions in some aspects of other services such as mental health related legal services, family law and criminal law. (Lattanzio, 2019, para. 2)

In 2023 the Ontario government brought in a $204-billion budget, a historical record. Much of this has been targeted towards major infrastructure projects such as building new highways and hospitals. Critics point out, however, that there is little or no assistance provided to citizens who are facing a cost-of-living crisis (Southern, 2023). Moreover, the government investment in infrastructure such as hospitals appears to be part of a controversial plan to further privatize the province's health care system. This will accentuate class divisions between those who can and cannot afford privatized care.

Under such austerity regimes, disabled people are among the most dispossessed people in our societies: constructed as citizens who "don't count," having already meagre benefits cut, and whose lives are placed in jeopardy. At the risk of sounding melodramatic, they and other vulnerable groups such as the elderly are increasingly becoming the victims of violent cleansing, albeit often invisibly so. The lesson is that austerity kills, although this is seldom acknowledged in our societies. One exception is Steinstra's (2017) feminist analysis of austerity and its implications for disabled women in Canada. In her article, she pointed out that disabled women, through especially low incomes combined with unemployment and cuts in government support and services, are particularly vulnerable to being coerced by the material conditions of their lives into seeking medical assistance in dying (MAID).

Grand'Maison and Lafuente (2022) tackled the urgent challenge of what they termed "dys-feminicide" or the murder of disabled women and girls who are viewed as "disposable" and less deserving of justice. Such acts are sometimes constructed as being a facet of care – for instance, as a way of ending a woman's suffering. Such constructions contribute to rendering such acts less visible than they might otherwise be. The authors called out for action on this issue:

> Finally, and most importantly, global feminisms risk replicating and reinforcing ableist notions of which lives are worthy of justice by erasing disability as one of the central elements of many women's lives. We urge feminists around the globe to reflect on the role of disability as an organizing principle that leads to the disposability and murder of women with disabilities. A tangible first step would be to systematically report the victim's disability status in feminicide observatories around the world. (p. 141)

Puar (2017) suggested that accounts of the uneven development of impairment and disability in our world need to consider "the right to maim" as a biopolitical force and set of practices employed by some able-bodied citizens as a way of asserting power over the lives and living conditions of others. Using the example of the Israeli-Palestinian conflict, she noted how Israeli soldiers at least sometimes exercise this right as a way of avoiding killings that would result in high civilian Palestinian casualty rates. So, for example, Israeli soldiers often shoot Palestinian people in the legs or head rather than the torso knowing that they will still be debilitated and face the harsh consequences of being impaired and disabled. This right to maim extends to vital infra-structures such as hospitals. Puar noted that the ongoing destruction of Palestinian infrastructure and repeated rebuilding has had especially dire consequences for those people who are "maimed" but at the same time has been a lucrative kind of "disaster" capitalism. She wrote:

> [D]ebilitation is extremely profitable economically and ideologically for Israel's settler colonial regime. Many sectors take on the "rehabilitation" of Gaza in the aftermath of war: Israel, Egypt, the Arab Gulf states [and] NGO actors who are embedded in corporate economies of humanitarian-ism. (Puar, 2017, p. 145)

Berlant (2007), a feminist scholar, uses the concept of "slow death" to capture "the physical wearing out of a population in a way that points to its deterioration as a defining condition of its experience and histor-ical existence" (p. 754). Slow death is a gradual wearing out of embod-ied subjects through the work of reproducing life. This certainly and tragically is the case in Palestine. Berlant continued:

> Slow death prospers not in traumatic events, as discrete time-framed phenomena like military encounters and genocide can appear to do, but in temporally labile environments whose contours in time and space are often identified with the presentness of ordinariness itself, that domain of living in which everyday activity, memory, needs, desires; and diverse temporalities and horizons of the taken-for-granted are brought into prox-imity and lived through. (p. 754)

The concept of slow death is useful in understanding the processes of oppression and dispossession to which people are subjected in an able neo-liberal capitalist global order. Even those who are deemed able face growing pressure to outperform co-workers and to survive in precari-ous, low paid, non-standard (e.g., involuntary part-time) jobs, including by juggling multiple ones (see May, 2019, for discussion of the Cana-dian context). Some workers even resort to performance-enhancing drugs to get the "upper hand" in the workplace in terms of productiv-ity (Brasheare, 2013). The stress and uncertainty associated with this situation and with austerity make even able-bodied and able-minded citizens vulnerable to "slow death." Having said that, the situation is even worse for groups such as seniors, women, and youth. May (2019) stressed this in the federal report on precarious work in Canada:

> Recently released Statistics Canada research documents that youth, seniors and workers without post-secondary education are more likely to be working part-time or temporary jobs. This may be either by choice or as the result of difficulty accessing standard employment. In 2016, about one-third of young Canadian workers were in part-time jobs, while this was the case for one-tenth of prime-age workers (ages 25 to 54). High rates of part-time work were also observed among seniors (31%), women (26%) and workers without post-secondary education (25%). Overall, their risks of falling into the poorer job quality classes were notably high: about 38% for women and for 18–29 year olds, and 34% for seniors and workers without postsecondary education. (p. 20)

An expert witness to the committee proceedings on which the report was based stressed that workers in precarious jobs who are injured or become ill at work lack adequate compensation and supports (May, 2019). Those who are injured or become ill in the course of employment are offered economic and rehabilitation supports that reflect their salary base. This means that part-time and temporary employees receive low levels of compensation and rehabilitation supports. In addition, rights to return to pre-injury jobs are frequently offered only to permanent employees. Precarious workers are thus disadvantaged not only by lower benefits but also fewer rights to return to where they were working prior to injury or illness (May, 2019).

Another witness commented on the fact that working in precarious jobs caused stress that negatively impacted her health:

> Allyson Schmidt also contributed to the Committee's understanding of the health impacts of insecure employment. She described the "stress of constantly applying for contracts and the stress of never feeling good enough or qualified enough." She also emphasized the negative impacts of insecure employment on relationships vital to good mental health, saying "[i]t is a big struggle to engage in the community." (May, 2019, p. 38)

Perhaps more dramatically, military conflict also contributes to the geographically uneven development of impairment and disability. In the case of the Palestinian people discussed by Puar (2017), Israeli soldiers exercising the "right to maim" oppress by instilling pain, intimidation, and fear and dispossessing people of able bodies and/or minds. As mentioned above, the destruction of the infrastructure helps to ensure that they live lives of "desperation" and have little chance of diminishing this without assistance in areas such as health care and prosthetic devices. This is a population, then, which is especially at risk of "slow death."

Garland-Thomson (2011) proposes using a feminist materialist concept of "misfitting" to understand how various anomalous bodies (e.g., disabled, raced, queer) are disciplined and excluded in capitalist society and space. The term signals that misfitting is a common experience for those who deviate from able, racial, heterosexual, white, male bodies and minds and thus has implications for political alliances or solidarity. So, for example, bodies that are large or "fat" experience misfitting in standard seating, for instance, in airplanes or in small change rooms in clothing stores (Longhurst, 2010). Disabled persons also experience misfitting in multiple ways: when stairs do not allow access to a building, and when they are subjected to invasive questions about their bodies, minds, and lives. Persons who identify as "queer" experience misfitting when court proceedings fail to allow for sexual identities other than hetero- or homosexual (Bains & Nash, 2007) or when transgendered teachers experience intolerance and hostility from students (e.g., Doan, 2010). While I make some reference to misfitting in the chapters that follow, I would also argue that this is a relatively weak concept in terms of helping us to explain why misfitting is a common occurrence in capitalist society and space. Its strength, then, lies in the basis it provides for political solidarity among a range of people with anomalous bodies and minds.

Perhaps not surprisingly, given the upsurge in right-wing populism and increasing disparities in economic well-being today, disability scholars and geographers are showing increasing concerns to further develop political-economic/Marxist accounts of ableness, impairment, and disability. Campbell (2009), for example, showed how deaf people were pressured, for profit-making reasons, to become "able" to hear using cochlear implants – even if they did not want this invasive aid and procedure.

Erevelles (2011) explicitly called for historical materialist analyses of disability that consider class relations as part of the process of becoming disabled in a capitalist world. She has argued that class has seldom been the focus of work in disability studies and yet, to understand the historical and material conditions of disabled people's lives, their class locations are vital to consider.

Other scholars, such as Goodley (2014) and Mitchell (2015), have been broadly sympathetic to this view. The latter drew on the concept of able-nationalism (defined as where ability and nationalism intersect) and Puar's (2017) notion that in contemporary neo-liberal capitalist societies more and more embodied subjects are being considered as "debilitated" and thus ripe for the consumption of commodities (e.g., hearing aids, pharmaceuticals, mobility scooters) that promise to make them more "able," once again, albeit often imperfectly so.

Erevelles (2011) contended that "humanist" constructions of disabled persons have inadvertently deflected attention away from the historical and material conditions of exploitation and violence that have helped to produce impairment and disability. Challenging tendencies in the disability studies literature to posit a non-racialized disabled subject, Erevelles (2011) has observed:

> [W]hat is left unquestioned are the historical and economic conditions that situate becoming disabled in a violent context of social and economic exploitation that may inhibit as well as complicate oppositional/transgressive theorizations of disabled subjectivity. It is to this fraught context, where be-coming disabled is produced within the actual material violence of transnational capitalism that I turn to in the next section. (p. 38)

Arguing that it is important to consider the conditions under which bodies become both racialized and disabled, she continued:

> I want to stress here that my argument is not that disability is like race or that race is like disability. Rather, I am arguing that within the specific transnational conditions of colonialism/neo-colonialism, the becoming of black disabled bodies is indeed an intercorporeal phenomenon that foregrounds a violent hierarchical context that contemporary theorists of difference have been reluctant to address or even acknowledge. (p. 39)

Examples of such violent hierarchical contexts include the slave trade and contemporary police violence against persons of colour – violence that kills or seriously injures – in countries such as Canada and the United States (for a discussion in the context of Toronto, Canada, see Hayes, 2018). I keep these contexts in mind, particularly in chapters 5, 6, 7, and 8.

Schalk (2022) has called for greater attention to what she terms "black disability politics." Drawing on historical accounts of organizing through the Black Panther Party (BPP) and the Black Women's Health Project (BWHP), both in the US, she argued that this has differed from mainstream white disability organizing in important ways. One is that it recognized, holistically, that disability is always experienced at the intersection of multiple oppressions – including those of race, gender, and class. So, for example, Black people are disproportionately likely to experience environmental racism (e.g., exposure to toxic substances in their communities) and incarceration in prisons or mental health facilities where "care" practices such as the psycho-surgery that the BPP strongly opposed, contributed to impairment, and disability.

As Schalk's (2022) work on BWHP helped to demonstrate, both "black disability politics" and "feminist black disability politics" recognized the importance of providing training and support to help their members articulate their demands and organize successfully despite the multiple oppressions they experienced. Schalk went on to show how today's Black cultural workers have built on these legacies to develop a movement for social change that explicitly recognizes that it is imperative to challenge a whole system of oppressions in global capitalism. Piepzna-Samarasinha (2022) made a compelling case for doing so through greater reflection on how diverse crip activists are imagining disabled futures and developing those futures through mutual aid practices that differ from white able aid – for instance, in terms of recognizing and celebrating the creativity that diverse disabled activists are bringing to the project of trying to ensure that disabled people are able to survive in their communities despite harsh realities such as insufficient access to food and shelter.

Along with calls for greater attention to disability in the Global South, there has been a growing recognition of the importance of situating our understanding of impairment, disability, and ableness in a neo-colonial global context. Scholars such as Meekosha (2011), Soldatic (2013), and Soldatic and Grech (2014) have argued that neo-colonial relations of power are central in understanding the oppression and marginalization of disabled people, particularly, but by no means exclusively, in the context of the Global South where most (80 per cent) disabled people live. Scholars such as Erevelles (2011) and Soldatic (2013) have suggested, respectively, that a transnational lens is helpful in bringing such relations and processes to light and in thinking about the transnational pursuit of claims for justice and reparations because of the production of impairment. Soldatic (2013) usefully pointed out, however, that there is a disjuncture between the transnational pursuit of rights for disabled people, particularly through the UN Convention on the Rights of Persons with Disabilities (CRPD), and the national context in which these rights are enforced.

Soldatic (2013) took the debate a step further and suggested that, at present, discourses and practices associated with the international pursuit of disabled people's human rights also play an important role in constructing disability as a universal experience and erasing related phenomena such as the production of impairment. The latter is, I argue, a geographically uneven process that concentrates impairment and disability in poorer nations of the Global South – those nations least likely to have the human and other resources needed to prevent or diminish impairment and to contest disabling barriers to inclusion. Clearly, such post-colonial outcomes constitute grave social and spatial injustices in the global capitalist order.

Geographers are making important contributions to our understanding of impairment and disability but have had relatively little to say about ableness. In fact, a 2020 search for ableness and ableism in the journals *Social and Cultural Geography*, *Progress in Human Geography*, and *Gender, Place and Culture* produced no results. Eighty-five articles on disability were found in the first journal but no references to ableness or ableism. A search of *Progress in Human Geography* found six articles on disability and eighty-one articles referring to disabled, but no references to ableness or ableism. The search of the journal *Gender, Place and Culture* revealed only one article referring to disability and eighty-one to disabled, but no articles referring to ableness or ableism. This is worrying, perhaps especially in these pandemic times, when some of our so-called "leaders" have prioritized reopening economies, profit-making, and the accumulation of wealth over the health, abilities, and lives of their citizens (Glasbeek, 2020). The same search in May of 2023 found only two references to ableist or ableism in the first journal, no references to them in *Progress in Human Geography*, and only two articles in *Gender, Place and Culture*.

In the next subsection of this chapter, I consider some of the contributions that geographers have made to our understanding of disability and argue that the time is ripe for a radical geography of ableness, impairment, and disability that challenges misconceptions that returning to the destructive and divisive capitalist status quo or "business as usual" is a viable option.

Thinking Radically and Geographically About an Ableist World

In recent decades, geographers drawing on social theory to address social issues such as disability have come to describe themselves as "critical social geographers." This is meant to convey, at least in part, a shift away from earlier radical geographies, eventually drawing primarily on the work of Marx, towards a body of work using a wider range of social theories to understand society and space. While such breadth is welcome, it also arguably poses risks of making a critique of capitalism less central to our work as scholars and activists. This is at a time when fault lines in our global capitalist order have deepened. This was evident in the rationing of health care (e.g., ventilators) to younger and more able patients during the pandemic, austerity measures threatening the lives of those most in need, and the ongoing polarization of wealth. When health care, social support, and access to economic wealth falter in such ways, disadvantaged groups (e.g., disabled people, racial minorities, seniors) are increasingly caught up in harrowing struggles to survive. Clearly, as developments like these help to remind us, capitalism only works for a wealthy elite – a broken system in which people die a "slow death" or, as Tyner (2016) put it, are "left to die."

The 1990s proved to be a decade of debate and reflection among geographers concerned with disability issues. Chouinard and Grant (1995) drew attention to the invisibility of disabled and lesbian women in the geographic literature concerned with oppression and resistance. Others critiqued behavioural approaches to understanding disability, notably the work of Reginald Golledge (R.E. Butler 1994; Gleeson, 1996; Golledge, 1996; see also Imrie, 1996). By 1999, Gleeson (1996) had published his historical materialist account of disability and the shift from feudal to capitalist societies. In the same year, R.E. Butler and Parr (1999) published *Mind and Body Spaces*, the first collection to share geographic research concerned with mental illness and issues of physical impairment. The book contained chapters dealing with topics such as personal experiences of chronic illness, barriers to disabled women's activism in both the disability and women's movements, and the impact of visual impairment on people's abilities to navigate the gay scene. I won't go into further detail on these early decades of geographic enquiry as they have been covered elsewhere (e.g., Chouinard et al., 2010; Park et al., 1998).

Over the past two decades, geographers have built upon the legacies of this work to advance our understanding of impairment and disability. Research undertaken in the 2000s was showcased in the 2010 edited collection entitled *Towards Enabling Geographies* (Chouinard et al., 2010). Among the changes in geographic approaches to disability captured in this collection were a broadening of the category of disability to encompass more embodied differences such as grappling with being a large or "fat" woman or being of short stature (Kruse, 2010; Longhurst, 2010). Kruse's (2010) and Longhurst's (2010) chapters identified a "mis-fit" between the embodied subject and her/his environment. In the case of "fat" women, for instance, scrutiny of bodies by others was a deterrent to using places such as the beach. Links between impairment, illness, and disability and changing experiences of everyday spaces of life also received attention. Imrie (2010) examined how the design of the home rarely allowed for the possibility that impaired corporeality might be a facet of life in the home and found that this resulted in physical and social barriers to the inclusion and well-being of disabled inhabitants. Crooks (2010) explored how women with the contested illness of fibromyalgia found their experiences of the home changing as, for example, some adjusted to fatigue, pain, and diminished interaction with others by making "high traffic" areas more central to their spaces of daily life. In a different vein, Metzel (2010) considered the social and spatial challenges of being a sibling caregiver at a long distance from their disabled sibling.

The years since 2010 have seen many exciting developments in geographies of impairment and disability. Here, I can draw attention to just a few examples. Drawing on neo-colonial theory, there have been efforts in and outside of geography to better understand impairment and disability in countries of the Global South (e.g., Chouinard, 2014; Ghai, 2013; Grech & Soldatic, 2016). Among the insights provided by such work have been the links between colonial and neo-colonial relations of power and violence resulting in the acquisition of impairments by women, and in violence against disabled women and men.

Paid work is often viewed as a way of promoting the inclusion of disabled people – not to mention as a foundation for austerity measures such as curbing social assistance to disabled people. But despite efforts to address diversity and inclusion in mainstream workplaces, there are many disabled people who are marginalized during hiring process and in relation to pay and opportunities in the workplace. This is one reason why Buhariwala et al. (2015) have explored the extent to which social enterprises may afford disabled people greater inclusion in the workplace. Marquis et al. (2016) have delved into ways of making universities more inclusive of disabled students. Wilton et al. (2013) have laid the groundwork for understanding the role of masculine identities in recovery from addictions. Among the insights from this latter area of research is that masculinity often needs to be reworked away from "macho" forms in the recovery process. Hamraie (2017) has examined the political economy of universal design in the United States. One of the insights her work has provided is how discourses about the benefits of universal design tend to obscure pressing issues such as the need for profitability and who wins and who loses from urban change. She wrote:

Another way of saying that history repeats itself is to point out that struggles for access, belonging and citizenship are not contained in a linear march toward progress. The consumer-centric post-ADA narrative that dominates much of Universal Design marketing tells us little about the sociopolitical economy of design or what purpose profitability serves, who benefits, and towards what ends. Buttressing this narrative, rehabilitation experts continue to raise alarm about the aging, white, suburban baby boomer population and its unique vulnerabilities in single-family homes and spatially dispersed, car-centric communities. Emphasizing the added value of Universal Design for this population, designers and manufacturers introduce new products into private consumer markets and attest to their commonsense, good design. It is easy to forget, however, that concerns about population aging have been raised for nearly a century, since even before today's baby boomers were born, that access to aging is a marker of privilege in a dominant culture that does not anticipate or value all lives. (pp. 257–8)

The challenges facing children caring for parents with HIV/AIDS and the forces shaping their caregiving practices have been explored in the work of Evans and Becker (2009). Drawing on research in the UK and Tanzania, they proposed an analytic approach which considered causal factors at different geographic scales: ranging from the individual and local to the global. Their work also contributed to efforts to address such issues in the Global South as well as the Global North. Among the insights offered by this work is the importance of considering global forces shaping young caregivers' lives and how this relates to the way their nations are situated in the global order. On this, the authors argued that economic restructuring, the HIV/AIDS epidemic, and growing disparities between rich and poor (e.g., in terms of life chances) as well as the outmigration of people from the Global South to the Global North (e.g., losses of trained medical professionals) have meant that low-income women and children are increasingly likely to care for household members with HIV/AIDS. As feminists have stressed, neo-liberal economic policies fail to take this unpaid reproductive work into account. Moreover, cuts in public health services have forced growing reliance on private resources to provide paid and unpaid care in the home. Evans and Becker (2009) described the material conditions of life shaping children's experiences of providing unpaid care:

> Evidence from both the North and increasingly the South suggests that children who become carers usually live in low-income families, where parents are unable to access whatever care provision is available, are unable to afford private healthcare or pay for homecare services or domestic labour (Becker et al., 2001; Robson, 2004; Becker and Becker, 2008). Indeed, Robson (2004, 227) suggests that young people's caregiving is "a largely hidden and unappreciated aspect of national economies, which is growing as an outcome of conservative macroeconomic policies and the HIV and AIDS explosion." (p. 3)

Other innovative research paths are being carved out by scholars such as E. Hall (2019) who has argued for greater attention to the wide range of acts of abjection that are directed against disabled people in various spaces of everyday life. He noted that while many people think about disability hate crimes as extreme acts of violence, many more are more mundane micro-aggressions such as name-calling, taunting, and harassment. Building on his work with Wilton (E. Hall & Wilton, 2018), he argued that non-representational theory, which emphasizes moments of encounter between subjects, objects, and spaces of life, can encourage a more relational, dynamic understanding of how able-bodied people interact with disabled people. One of the examples that he gave was that sometimes when disabled people at malls use disabled parking spaces they encounter anger from and name-calling by able-bodied people who do not believe they are entitled to use designated parking spots.

One encounter of this sort that I had was when I had just parked in a disabled parking space and a woman parked her car directly behind my vehicle and began shouting. It was only when I got out of the car with difficulty using a cane and got my walker out of my trunk that she went silent and drove away. Such encounters are distressing and certainly make one feel unwelcome in spaces such as the mall – spaces primarily designed for persons with able bodies.

While much of this work by geographers has been at the local level or on a micro-scale, there are also examples of research that is national or even transnational in scope. Wilton and Evans (2016), for example, considered the potential of social enterprises in Canada as places that can encourage more inclusive encounters between disabled and non-disabled people. In my own work on disability in the Global South (specifically Guyana), I emphasized how important our location in the world is to our experiences of impairment and disability. How, for example, living in a poorer nation such as Guyana can result in heightened barriers to receiving health care or getting prostheses (because of limited government funding and the outmigration of health professionals and other factors such as lack of public transit services) (Chouinard, 2014, 2019). In the chapters that follow I am mindful that experiences of impairment, disability, and ableism are embedded in the specific context of places existing at various geographic scales in the Global North and the Global South.

Summary: Towards a Radical Geography of Ableness, Impairment, and Disability?

In this chapter, I have argued for a materialist, feminist, queer, political-economic, neo-colonial, and geographic conceptualization of processes reinforcing ableness and the disabling of persons with impairments and illnesses. This seems to me to be essential at a time when able neo-liberal capitalism has helped to unleash forces such as right-wing populism that are stoking intolerance towards persons who embody differences such as impairment, race, gender, and class. Among the signs of this are an American ex-president who has openly mocked a disabled journalist and made sexist and disparaging comments about women, people of colour, journalists, and many others.

In conclusion, I want to focus on the implications that the ideas discussed in this chapter have for thinking about ableness, impairment, and disability as matters of social injustice. Ableness is, to use a broad definition, a regime of power and privilege that favours those who can most closely embody having an able body and mind and disadvantages those less able to do so. It is a regime and set of practices that reinforces "compulsory able-bodiedness" as well as the aspirational ideal of the able citizen who progresses, for instance with respect to health and wealth, as part of a meritocracy. Fundamentally, then, ableness is about rewarding those who are fortunate to have able bodies and minds and penalizing those who are not. Those rewards include better economic prospects, higher incomes, and greater access to education. Penalties for not being able include poor economic prospects, exclusion from places of paid work, and representations of non-able lives as being "not worth living." In terms of dynamics, then, ableness is about perpetuating social injustices rooted in who does and does not enjoy the privileges of ableness. This compounding of privileges arguably makes as little sense as forcing people to sit at the back of the bus because their skin isn't white or singling out LGBTQ+ people for violence and harassment because they defy dominant gender norms. They are all unjust social practices rooted in intolerance towards diversity in embodied life. This includes limited access to spaces of life that able people often take for granted – for instance workplaces.

It is important to recognize, as Wolbring (2008) has suggested, that ableism is a regime of privilege and power that, in our neo-liberal times, values some abilities over others. The abilities to compete and to amass wealth, for example, are valued over other abilities such as the ability to show compassion and empathy for other human beings and beings more generally.

One way in which these injustices find expression at the national and transnational scales is through an "able-nationalism" (Mitchell, 2015) manifest in terms of who counts as a valuable citizen and in related practices such as denying immigrant status to persons deemed an undue burden on social and health care systems (on this in the Canadian context, see Wilton et al., 2017).

While I concur with scholars such as Erevelles (2011), Meekosha (2011), and Soldatic and Grech (2014) on the need for transnational approaches to disability issues as matters of social injustice, I also agree that international human rights instruments such as the CRPD obscure as much as they reveal through a universalistic construction of disability as a generic experience. An important example of this is a pervasive erasure of issues such as the geographically uneven production of impairment that characterizes our contemporary able global capitalist order. As some disability scholars have begun to recognize, this, along with ableism, must be part of transnational struggles for greater justice and inclusion for the over one billion disabled people on our planet.

Amid a global health and economic crisis that is revealing class and other disparities or "fault lines" in our global capitalist system, I think it is worth asking whether the time is right to develop a radical geography of ableness, disability, and impairment. By this I do not mean returning exclusively to Marxist theory. What I mean is that, while we can maintain or enhance the breadth of theoretical frameworks that informs current disability studies, we can also insist on the radicalness of this endeavour. If radicalness means getting to the root causes of things, including disparities in power, then we need to address how ableness, disability, and the production of impairment are rooted or embedded in the able global capitalist world we find ourselves in.

We also need to commit to finding ways to challenge a global order that is leaving so many people to die in the name of profit and extreme wealth. As the global pandemic and rush to reopen economies, austerity policies, precarious work, rising costs of living, and extreme weather events all continue to claim lives, they are also helping to lay bare just how many people do not count in our world and are being condemned to slow and not-so-slow deaths in the name of profit and wealth. I believe that we can no longer afford such a destructive, violent, and unjust system. If this book contributes to thinking more radically about how our ways of life need to change, it will have been worthwhile.

In the next two chapters I illustrate facets of these dynamics of ableness, impairment, and disability at the micro-scales of the intrapersonal and interpersonal: first with respect to growing up in ableist and misogynistic environments in the home and community, next with respect to being a disabled female professor in an academic workplace, and finally being a disabled woman in the wider community. The following chapter continues to focus on the micro-intrapersonal and interpersonal scales but this time with respect to what it is like to negotiate bipolar "madness" in places of everyday life. Here I draw on my own experiences of this illness as well as on autobiographic accounts by others.

3 On Living in an Ableist World (without Necessarily Knowing It) and Being Disabled in and outside the Academy: Autoethnographic Reflections

Over the last almost three decades, my concern with disabled people's lives has been fuelled, in part, by my own personal experiences of being disabled in the academic workplace and other spaces of life. It therefore seems fitting that the third chapter of this book use the personal as a window into issues of disability and ableism. However, and unlike my previous reflections of this type, I am concerned here with expanding on these by considering my experiences of ableism during various stages of my life and in non-academic as well as academic spaces. By expanding the political and spatial scope of my autoethnographic reflections and making the personal political/cultural, I hope to further highlight how disability and ableism help to shape our lives in oppressive and exclusionary ways.

The chapter begins with me sharing how I understand disability and ableism or ableness. This is followed by an overview of what we do and don't know about ableism and disability in and outside the academy – given the now extensive literature on geographies of disability and disability studies more generally, this is a necessarily selective overview. I have, however, tried to zero in on what I see as some of the most significant lessons we've learned. This is followed by some retrospective autoethnographic reflections on childhood and what it was like to grow up in a profoundly ableist and misogynist world even if I was relatively oblivious to it at the time. Next, I consider some of what can be learned about this world from my experiences of being a female disabled professor and a disabled woman more generally. Then I share autoethnographic reflections from my current vantage point of being a senior disability studies scholar and geographer. In conclusion, I suggest some of the reasons why it is so important that we have further autoethnographic accounts of ableism and disability.

Understanding Disability and Ableism

As noted in chapter 1, disability is understood not only as an outcome of environmental, attitudinal, and cultural barriers to the inclusion of people with impairments and/or illnesses in society and space but also as an embodied and relational process of becoming. This avoids drawing a misleading dichotomy between mental and physical impairment as medicalized bodily states and disability as social (as in the traditional social model of disability). It also allows us to consider aspects of impairment and/or illness such as chronic pain as part of the experience of being a disabled child, woman, or man. Moreover, by thinking about disability as a relational process of becoming, we can shed light on how interactions among disabled people and encounters between disabled and non-disabled people in different places in the world help to shape conditions of daily life as more and less disabling. We can also consider how interactions with non-human others such as service dogs and with objects such as prostheses or other aids work in geographically uneven ways to enable or disable persons with impairments and/or illnesses. So, for example, my work on disabled people's lives in Guyana showed how being in the Global South severely limits access even to aids such as wheelchairs, mental health services, prosthetic legs and arms, and the technicians needed to design and fit them.

Despite being pervasive in our contemporary world, ableism is arguably trickier to define. This is at least in part because, as a regime of power and privilege that values able minds and bodies over disabled ones, our experiences of ableism tend to be a taken-for-granted aspect of our daily lives and of the privileges that accrue to those who can, at least temporarily, most closely approximate the able ideal. Such privileges include greater access to higher education, higher incomes, better employment conditions, and better homes, as well as being regarded as "successful" and "productive" members of society. Whether or not we have able minds or bodies, we are all engaged, albeit sometimes in different ways, with political, economic, and cultural practices of ableist power and privilege. My autoethnographic reflections in this chapter are meant to help illustrate this.

What Is Autoethnography?

In recent decades, autoethnography has emerged as a powerful qualitative research approach. Inspired in part by feminist scholars who emphasized that personal experiences have wider political implications (i.e., "the personal is political"), it is an approach that combines autobiographical writing with ethnographic reflections on what personal experiences reveal about the cultural workings of society. Or as Ellis et al. (2011) have put it, "Autoethnography is an approach to research and writing that seeks to describe and systematically analyze personal experience in order to understand cultural experience" (p. 1). The crucial point here is that personal stories are not told for their own sake but because they reveal something about the cultural dynamics of the societies and spaces in which we live. Autoethnography also resonates with the radical view that the point of knowledge is not merely to understand the world but to change it for the better. Ellis et al. (2011) have noted in the opening of their article that "autoethnographers view research and writing as socially-just acts; rather than a preoccupation with accuracy, the goal is to produce analytical, accessible texts that change us and the world we live in for the better" (p. 1).

The autoethnographic reflections shared in this chapter differ in important ways from my previous writings of this type (Cameron & Chouinard, 2014; Chouinard, 1995/1996, 2010). This work has focused primarily on my experiences of gendered and ableist oppression and disability in the academy. Here, I aim to provide a more expansive understanding of negotiating ableism and disability at different stages of the life course and in a range of spaces of everyday life. I would argue that such accounts are vital in helping us to better situate experiences of impairment, illness, and disability in the context of an oppressive able neo-liberal capitalist world (see also Erevelles, 2011; Goodley, 2011). Or, to put it slightly differently, we can better understand how ableness is lived by diverse embodied subjects in today's global capitalist order.

In the next section, I provide an overview of what we do and still don't know about experiences of ableism and disability in and outside the academy. This is followed by some retrospective autoethnographic reflections on childhood and what it was like to grow up in a profoundly ableist and misogynist world even if I was relatively oblivious to it at the time. Next, I consider some of what can be learned about this world from my experiences of being a female disabled professor and a disabled woman more generally. Then I share autoethnographic reflections from my current vantage point of being a senior disability studies scholar and geographer. In conclusion, I suggest some of the reasons why it is so important that we have further autoethnographic accounts of ableism and disability.

What We Do and Don't Yet Know about Ableism and Disability in and outside the Academy

You may recall the old adage "the more we learn, the less we know." This certainly is an apt observation regarding geographic studies concerned with disability and ableism and disability studies more generally.

As noted in the previous chapter, over the past three decades scholars have made some exciting contributions to our understanding of disability and, albeit to a lesser extent, ableism in society and space (e.g., Campbell, 2009; Longhurst, 2010). In my home discipline, geography, the number of geographers working on disability issues has grown. This is a welcome development. However, it is prudent, if potentially uncomfortable, to ask who these geographers are and whether and if so, how, they are working in solidarity with the relatively few disabled geographers represented in the discipline.

Like the broader discipline, many geographers who study disability are white, male, and at least visibly able-bodied (which may be a result of fear of disclosure in an ableist academy or able-bodied status or a combination of both). There is also at least anecdotal evidence that geographic studies of disability conducted by scholars who are themselves disabled are at least sometimes marginalized in the discipline.

For example, I was involved virtually in two panel sessions organized by the Disability Specialty Group (DSG) of the American Association of Geographers in 2018. (My attendance was virtual since I can no longer travel on my own due to mobility impairment and pain and because the chair of this specialty group was open to doing things such as conferences differently if it heightened inclusion of disabled scholars.) What was disturbing, however, was her observation that there did not seem to be any able-bodied geographers in attendance at DSG sessions. If so, this arguably undermines disability scholars' and activists' demands that there be "nothing about us without us" (Charlton, 1998) and is a worrying sign that able-bodied geographers are at least sometimes not working in solidarity with their disabled counterparts or at least not working as closely as they could, or should. I raise these issues not to disparage the excellent work that many able geographers do, but to suggest that there are important issues of exclusion in the academy that are crying out for more attention and debate over the short and longer terms. One of the more pressing of these issues is what scholars are prepared to do to ensure that more disabled scholars are represented in their disciplines. This is important in practical ways such as mentorship and in terms of conscious efforts to work in political solidarity with disability scholars who might not otherwise have a chance to be included in the academy.

As I write these words, I am conscious that when I leave my academic unit, it is unlikely that a disabled scholar will replace me. I say this because, in part, it underscores how exclusionary the academy can be and because, in part, it has been conveyed to me in numerous ways over the years that my "slow scholarship" and failure to match the output of my able colleagues means that I have "failed." I would argue, however, that it is the academy that has at least in some ways failed me by, for instance, not taking the different career trajectories of disabled scholars into account in decisions on matters such as salary and performance. I think any serious challenges to the exclusionary academy must take such acts of ableness into account and I hope we will see this happen in the not-too-distant future.

In this context, it is perhaps not surprising that geographic studies of disability have had little to say about what it is like to be disabled in ableist academic settings. While there are some exceptions, such as Cameron and Chouinard (2014), Chouinard (2010), La Monica (2016), and Marquis et al. (2016), for the most part the voices of disabled scholars in the discipline have been silenced and marginalized. We can understand this, at least in part, by the fact that disabled people continue to be excluded from spaces of academic life. In fact, as I recently told a colleague, the only reason I continue to have a presence at my academic institution is that I became disabled after being appointed to a tenure track position. Inclusion should not, however, be a matter of "luck" or in my case also a prolonged and very draining struggle for accommodation and should be based on a recognition of how aspects of diversity such as disability enrich academic as well as non-academic environments.

So, what have we learned from geographers about ableism and disability in the academy? One important insight is that most of what we know pertains to undergraduate students living in more affluent nations (La Monica, 2016). There is thus a pressing need for research that considers ableism and disability in the lives of graduate students and professors, especially those living in countries of the Global South. Another insight is that personal struggles for accommodation in the academy can be protracted and harsh (e.g., Cameron & Chouinard, 2014; Chouinard, 2010; La Monica, 2016). For the most part, however, geographers have had little to say about ableism, disability, and the academy. This needs to change.

Thanks to work by disability scholars based in other disciplines, we also know that ableism is deeply entrenched in academic culture. For example, as Price (2011) has demonstrated, there is little space and tolerance for mentally disabled people in the academy, especially with regard to those with psychiatric illnesses. This should not be surprising since academia prizes what is perceived to be rational thought above all else (perhaps except for profit-making in these neo-liberal times). Instructors also, as Price (2011) has shown, engage in practices such as attributing a student's absence from class to a lack of motivation when it could also signal mental disability and distress that could make it impossible to even get out of bed let alone go to class. Dolmage (2017), who wrote on ableism in North American higher education, drew attention to the eugenic thinking that informed higher education in recent history and provided a justification for practices such as forced sterilization of persons with mental disabilities:

> When Margaret Price published "It Shouldn't Be So Hard" in *Inside Higher Ed* in 2011, arguing for a more accessible climate for disabled faculty, for example, numerous responses to the essay only reinforced the sense that faculty members should mask or carefully disguise any weakness or inability to perform ... Not much has changed in the intervening semesters since 2011. Despite some calls for faculty to be more open about disability, the administrative and cultural milieu for disabled faculty remains relatively inhospitable whether overtly or covertly. It is still very, very hard. (p. 177)

He continued: "Disabled people with PhDs are much more likely to end up under-employed, exploited as adjunct labor, and to experience discrimination as a result of their disability" (pp. 177–8).

Other disability studies scholars have pointed out that disability avoidance in higher education is important in shoring up and perpetuating an ableist academic order. Oswal (2018), in a critique of a 2012 report by the American Association of University Professors (AAUP) on how to (and not to) accommodate disabled faculty, noted that even the report's language is skewed towards the interests of maintaining an exclusionary able academic order. He observed, for example, that what is and is not regarded as a reasonable accommodation receives much more emphasis than diversity and inclusion:

> In this chapter, I trace the theme of avoidance in the academy as it occurs in policy formulation. In analysing the document, I show how a well-meant academic discourse about disability can degenerate into bureaucratic and legal quibbling about reasonable and unreasonable accommodations for disabled faculty, instead of creating an inclusive environment to welcome diversity through disability. By stressing the legal boundaries of accommodation, the AAUP document reduces discussion about inclusion to a list of dos and don'ts. Further, by avoiding the questions of social prejudice in the University, it fails to question the perpetuation of exclusionary, nondisabled policies and decision-making that would maintain the status quo and leave the ableist order undisturbed. (p. 21)

Other disability scholars, such as Titchkosky (2011), agree that disability studies must focus as much attention on how the academy treats persons with impairments and illnesses as on other issues. Titchkosky (2011) quoted Snyder and Mitchell (2006) who wrote: "Disability studies must recognize that its critique should be trained on the institution of the academy as much as on the social and political context outside its walls" (p. 70). For Titchkosky, this included analysis of the justificatory narratives used to try to legitimate unjust and exclusionary situations such as having inaccessible washrooms in her academic building but with misleading signs suggesting that they are accessible.

There are also general lessons to be drawn from the interdisciplinary disability studies literature that can help to guide how we go about understanding people's experiences of ableism and disability in and outside the academy. One, as scholars such as Davis (2017) have stressed, is how central notions of "normality" and "abnormality" are perpetuating ableist oppression and disabling conditions of life for those with impairments and illnesses. Davis noted the connection between the eugenics movement, with its belief in the potential perfectibility of the human race, and statistical sciences (e.g., psychology), as bases for measuring and classifying various characteristics of the human body and mind as "normal" and "abnormal" (e.g., in terms of intelligence or sanity/insanity, and in terms of physical mobility and so on). Challenging "normality" is difficult because it is an often taken-for-granted aspect of the ableist societies and spaces in which we live. However, as geographers Hansen and Philo (2007) pointed out, one possible strategy for doing so is to insist on the "normality" of doing things differently (e.g., wheeling as an alternative to walking). To the extent that we can reclaim "normality" for the ways in which disabled people navigate and assign meaning to their environments, we also help to disrupt the hegemony of the normal/abnormal way of thinking about people's abilities and lives. A concrete example of this that comes to mind is Gregor Wolbring, a disabled bioethicist who appears in the 2013 documentary film *Fixed*, which deals with human enhancement (e.g., bionic legs). Born without legs, he sometimes uses a wheelchair to navigate spaces. However, he does not use prosthetic limbs and several times in the film he also insists on the normality and even desirability of crawling (one memorable line from the film is: "crawling is in, walking is out") (Brashear, 2013).

Another important lesson from the disability studies literature is the need to think further about how neo-liberalism and ableism inform and perpetuate one another at scales ranging from the personal through to the global. Goodley (2014) described what he termed "neoliberal-ableism" in the following way:

> Disability is the quintessential Other of neoliberal-ableist society. Ableism edits out lack and emboldens (hyper) normality. There is no doubt that ableism is not only a matter of life but also of death. In our contemporary

times of austerity we are witnessing a reduction of state support alongside a reworking of the ableist ideals on independence and autonomy. This pincer action is having a huge impact on citizens of nation states: reduced potential support and expected to occupy an individualist responsibility for one's lot in life. Inevitably, there will be collateral damage. (p. 33)

In the autoethnographic reflections to follow, I draw on notions of normality and abnormality as well as the neo-liberal context of our lives to help make sense of the experiences I share.

But first I want to briefly consider some of the ways that feminist scholars have enriched our understanding of ableism and disability. Over the last three decades, feminist scholars have made important contributions to our thinking about disability and, to a lesser extent, ableism. One was to point out the neglect of disability issues and disabled women's lives in disciplines such as geography and philosophy (e.g., Chouinard & Grant, 1995; Wendell, 1996). Another was to insist on taking the body seriously in analyses of impairment, disability, and chronic illness (e.g., Dyck, 1995). More recently, there have been efforts to expand our understanding of the types of body differences that can be disabling (e.g., Longhurst, 2010). Feminist scholars have also helped to put intersectionality at the forefront of analyses of oppression and resistance. Put simply, intersectionality refers to how an individual is located at the crossroads of multiple groups. Valentine (2007) showed how the forms of oppression experienced by a lesbian D/deaf woman changed with the different spaces she negotiated and as aspects of her own life became more or less salient in how she was oppressed in different places of everyday life. Valentine demonstrated, for example, how being a lesbian woman became salient in the Deaf club she belonged to at the time because other members were hostile to her sexual orientation and relationship with another female member.

In the reflections that follow I will also illustrate how these contributions inform my assessment of my experiences of ableism and disability in and outside academia.

Growing Up in an Ableist and Misogynistic World (without Necessarily Knowing It)

Until recently, I did not associate my childhood years with growing up in an ableist and misogynistic world. After all, I was not disabled at the time and, although my father was sometimes verbally and physically abusive, I was too young to realize that this was an unacceptable way to treat a young girl. For my father, my mother, my child self, and my sister, my father's efforts to dominate and frighten us seemed "normal." It was only later, as I became a more critical teenager, that I realized this was not the case.

Over the years, as I have reflected on my experiences as a child, it has gradually become clearer just how ableist and misogynistic my world was. One of my recollections of being a young girl in public school in Canada was of being repeatedly taken out of my class for psychological tests to determine if I was intellectually "gifted."

Since I was the only student being taken out of class, I felt singled out as different, albeit in ways that teachers and psychologists seemed to regard in a positive light. When the results came back, they suggested that I was "borderline genius" and my parents were given the options of putting me in a special class or having me skip a grade. My parents, fearing that my intellectual abilities would not be adequate to allow me to compete with bona fide geniuses in the special class and that this would damage my confidence and performance, opted to have me skip a grade instead. These recollections help to remind me how pervasive ableism is in all our lives and that its workings are discernible even in the lives of young girls and boys. In a society that values able minds and bodies over others, there is a relentless drive to categorize and sort people according to ability and to how those abilities compare to those of others. It is a frenzied drive to determine who is best able to outperform others whether through psychological testing, athletic prowess, creative arts, and so on. Those who like myself "misfit" such embodied criteria often find themselves held back.

Competitive sports based in schools and communities are another realm where children and young people are drawn into ableist ways of thinking about their lives – again without necessarily knowing it. I still recall the trauma, in public school, of being among the last to be picked by team captains for baseball and field hockey games. Although I realized I was not gifted in my abilities to play these sports, I still felt inadequate, rejected, and singled out. Conversely, when I made it onto the school's volleyball team, I felt deep pride at my ability to compete in this sport. As I realize now, such feelings are also a part of living in an ableist world. We become emotionally invested in not having our abilities doubted and in accruing signs that we are able and even hyper-able (such as trophies, ribbons, membership on teams).

As a child, my experiences of ableism were complicated in significant ways by patriarchal oppression and the misogyny associated with it. My father was a large, imposing, and domineering man who lost his temper easily and was sometimes abusive. When angry or frustrated, he used his physical size and abilities to dominate, intimidate, and bully others – especially girls and women but sometimes men as well. I remember my mother recounting an incident when I was a toddler and did something "wrong." My father responded by spanking me so hard that there were bruised handprints on my bottom for a week. Hearing about this experience as a young girl was disturbing and, although I couldn't articulate it at the time, in retrospect it was clearly an example of using disparities in physical abilities to dominate and frighten me. This is arguably one of the uglier sides of living in an ableist and patriarchal world.

My father was not the only man I encountered who used disparities in size and ability to try to dominate and harm me. I recall, as a five-year-old girl walking to public school in the United States, when a man in a car stopped and opened the car door while masturbating and asking me, "Have you ever seen one of these before?" Fortunately, as he tried to coax me into the car, I recalled a warning that my parents had given a few days before that it wasn't safe to interact with strange men. I continued to walk to the school and, fortunately, he drove away.

As a teenager and young adult, ableism and patriarchal oppression continued to permeate my life and life spaces. As a teen still living at home, my father, as I've recounted in more detail elsewhere (Chouinard, 2014), continued to try to assert his authority by capitalizing on his physical size and abilities. This was often in situations where I had angered him by expressing an opinion on an event or issue that was not the same as his. Such actions elicited violent reactions from him, including slamming his fist on the dining-room table and shouting that he was "right" and I was "wrong." It made for tense meals, to say the least. But it also illustrates how ableist and patriarchal regimes of power and privilege intertwine and feed each other. My father was able to assert his sense of entitlement to adjudicate what was regarded as "true" in our family by virtue of the fact that he could use his physical size and abilities to dominate discussion as well as the fact that he was a man and considered himself "master of the home." There were also moments of verbal and emotional abuse, especially with respect to my decisions about dating and boyfriends. For me, the possibility of him also becoming physically abusive was always there, which made such actions especially disturbing.

In retrospect, it is clear to me that such experiences were intended, at least implicitly, to impair my ability as a young woman to gain confidence, independence, and skills for sound decision-making. This is another example of how ableism and patriarchy intertwine. In my father's case, a childhood that was difficult in some ways (e.g., a mother who was domineering and sometimes dishonest) had instilled an anger that lasted a lifetime, and he coped by lashing out at others, particularly at home, since he regarded this as his "turf."

High school was, of course, another realm for continuing to experience (but not learn about) ableism. Here the emphasis was increasingly on outperforming peers both intellectually and physically. The possibility of not being accepted for university was used to help instill a sense of urgency about doing well in terms of grades and other accomplishments (for example, excelling in a play or at sports). Not all students bought into this, of course, but it certainly signalled ableist attitudes towards students' performances in school. In my experience, there was also a sort of "social ableism" at work during high school and, later, university. By this I mean pressures to succeed at aligning oneself with the "popular crowd," something which tended to require physical attractiveness, trendy clothing, and an attitude of superiority. I vividly recall a new female student in one of my classes who approached me after class and asked, "So who are the popular students?" Despite not belonging to this clique, I was able to rattle off names and she happily sauntered off and did indeed become one of the "in crowd."

Although going away to university provided some badly needed relief from my father and former home, pressures to outperform my peers academically and to some extent socially intensified during my undergraduate days. For myself, and at least some of my peers, the new "normal" became to work very hard to demonstrate that we were "hyper-able" at our academic studies and work in a social life (not to mention sleep) when we could. As a young able adult student, it never occurred to me that responding to those pressures could be unhealthy and debilitating in the long term, and I recognize now that being oblivious to that was part of the ableist privilege I then enjoyed.

Pressures to demonstrate hyper-ability intensified as a graduate student, particularly at the PhD level. I had married just before beginning my master's program and, by the time I defended my PhD dissertation, I was a mother to a three-year-old daughter and a six-month-old son. This inevitably made it more difficult to continue to be the hyper-able academic I'd learned to take as being "the norm." However, this paled in comparison to the working conditions I encountered after accepting a tenure-track position at the university where I completed my doctorate, as I explain below.

Welcome to Academia (Well Sort Of)

Although I remember feeling bewildered at the working conditions I encountered as a new assistant professor, it is clear in retrospect that I was constructed as a negative "other" and not really belonging from day one. And that this lay at the heart of what others would later describe as a toxic or poisoned working environment. First, I was the "wrong" gender as all my colleagues were men. This made me a target for blatantly sexist and dismissive remarks. I recall a male colleague observing that "if my academic career didn't work out I could always stay home with the kids." Then there was the department chair, who had a reputation as a womanizer, and who subjected me to professional harassment. He would arrive unannounced in my office, often early Friday evening when others had left, and berate me for publishing in what he claimed were the "wrong" journals and for being less successful than my male counterparts. On one occasion, he berated me for working too hard, which shouldn't have been surprising as I had been given some of our very largest classes to teach, and inappropriately observed that this was "causing my looks to deteriorate." He also started to come into my office more frequently and hang around my desk, something I reported to security as soon as he left. There were also indirect forms of professional harassment, seemingly since I was the only faculty member doing "radical geography" at the time. I saw student after student in my office and they all said the same thing: that they wanted to work with me but had been told it wasn't a good idea because I "didn't do real research" and they'd never get a job.

As if these working conditions weren't toxic enough – and they were toxic enough to result, for example, in my being hospitalized with what would later be diagnosed as manic depression – my working life deteriorated further when it became clear, in 1990, that I was seriously ill with rheumatoid arthritis. The chair of the department told me in no uncertain terms that my career was over and that I should leave. I got the same message from several administrators. Disabled faculty were, it seemed, not only disposable but irrevocably "damaged goods" in the academic workplace.

When I resisted disappearing and having to rely on benefits because I enjoyed my academic work, I became embroiled in what would become a gruelling and prolonged struggle for accommodation as a disabled professor (as required under Canadian human rights law). The struggle began as an internal one in which the university staff working in the equity area and I tried, for eight years, to convince university administrators that basic accommodations made in 1994 (consisting in relief from undergraduate teaching in exchange for reducing my salary to one-half and requiring the second half to be paid as benefits from the insurer while permanently freezing my salary at 1990 levels) did not constitute "reasonable accommodation" and discriminated against me on the prohibited ground of disability. To our dismay, the administrators we dealt with were almost without exception hostile to the suggestion that the university had a duty to accommodate disabled professors and staff. Nor did they seem to understand, let alone care, that the duty to accommodate to what is termed "the point of undue hardship" (usually defined in terms of financial hardship but sometimes involving health and safety concerns) was something they were legally obliged to do. So, the administration came forward with proposals that continued to discriminate based on disability (readers wanting further details on these struggles can consult Chouinard 1995/1996, 2010). Put simply, their proposals would have enshrined discrimination in the handling of my case by the university. The provost, for instance, at one point proposed granting me a salary increase that reflected my promotion to full professor but enshrining a 50 per cent reduction in any earned salary increases (he argued that this was appropriate to reflect a "reduced workload" while we argued that, particularly after prolonged financial discrimination imposed by the frozen salary, this was just another discriminatory proposal).

In 1998, in exasperation, I hired a human rights lawyer, and began to take legal action against the university. I had to pay for this out of my own low salary because the university Faculty Association at the time refused to assist me (arguing that this was a "special case" and thus not relevant to other faculty members). This was preposterous in several ways, including that it ignored the fact that accommodation was by this point in time the right of disabled faculty and others with needs such as caring for children and/or parents under the university's own policies. But, instead of recognizing this, the Faculty Association claimed that what was "special" about my case was that it was about "early retirement" – a ridiculous claim since my case was about continuing to work with appropriate accommodations in place. It seemed at the time that the association was willing to "parrot" the position taken by senior administrators. The association, to its credit, had changed its position by the early 2000s and helped, along with the Canadian Association of University Teachers, to pressure the university into putting a new and less discriminatory accommodation plan in place. This did not happen until 2003, however, as even after signing a legal settlement with me in December 2000 that required the new plan, the university resisted putting it in place. Clearly, I continued to be viewed as damaged, disposable, lesser, abnormal, and all the various tropes deployed to devalue people with impairments and/or illnesses through an ableist gaze on one's world. In an ableist academic workplace, the bottom line was that the administration still wished I would just vanish.

Even in the early 2000s, administrators sometimes argued for discriminatory measures in the new academic accommodation plan.

The chair of my academic unit, for example, suggested that I be forced to reimburse the university, from my research funds, for any of the teaching costs associated with any undergraduate course that I would have taught had I not been disabled and shouldered the normal teaching load. My lawyer rightly objected that this was an attempt to make me rather than the university pay for being accommodated – something unheard of in the Canadian legal context. Such an arrangement was not only discriminatory but, in my view, also would have constituted a misuse of research funds. Fortunately, the proposal did not have sufficient support to end up as part of my new accommodation plan.

I would be remiss if I didn't mention that during this struggle for accommodation, that lasted for over a decade and a half, there were times and places where the intersectionality of gender and disability became especially salient in understanding my experiences of struggling for inclusion. This was often in situations where more powerful men were intent on wielding their authority in ways that, at least implicitly, sought to discipline and punish me as a woman acting on disability issues. These encounters almost always took place behind closed doors in administrators' offices – something that could have been seen as justified in terms of confidentiality but also could have been seen as male administrators being given more leeway to be abusive and bullying. As I've recounted elsewhere (Chouinard, 1995/1996, 2010), these incidents ranged from repeatedly attacking my character and behaviour (e.g., portraying me as insufficiently grateful for any accommodations I received) to a meeting with several senior administrators, the equity officer, and myself that was supposedly about the new accommodation plan but which rapidly degenerated into efforts, on the part of the Acting Provost, to humiliate and bully me and perhaps the equity officer as well. The Acting Provost did this by categorically rejecting each and every proposal we made about how I could contribute to undergraduate teaching. Most of the other administrators around the table, including one woman, remained silent even when it became clear that this treatment was distressing to me. After the meeting, I told the Acting Provost that a more constructive discussion of this matter was needed before leaving in disgust. I learned afterwards that other female professors active on different issues had similar experiences of "being punished" – clearly, this was a way of trying to assert male authority and power. And in my case, this also involved asserting able privilege in the academy. By portraying me as the ungrateful and defective female professor whose gender, impairment, disability, and abnormality meant she was the quintessential "other" of the able male academic ideal, they sought to buttress notions that women like me did not belong in the academy (and, of course, that they did). They were also asserting the patronizing ableist and patriarchal expectations that disabled female professors such as myself would express gratitude to them for any accommodations made as opposed to regarding such matters in terms of meeting the accommodation requirements of human rights law.

In the final section of this chapter, I will share autoethnographic reflections on gender, disability, and ableism in and outside the academy from my current vantage point as a senior feminist/social geographer and disability studies scholar. But first I share a few reflections on being a disabled woman in spaces outside the academy.

On Being a Disabled Woman outside the Academy

While most of my time and energy was spent battling ableism in the academy and dealing with physical impairments such as extreme joint pain, there were also experiences of disability and ableism in spaces outside the academy that are worth sharing.

One such space was that of the family. My partner and two young children were traumatized by events, such as my hospitalization, triggered in part by the stress of working in a hostile and toxic environment. This is a useful reminder that the consequences of being disabled and experiencing ableism as well as gender oppression in one space of life is not neatly "contained" there but seeps into other realms and spaces of life. There were also challenges associated with finding "non-able" ways of conducting family life. My partner and children had to accept my needs for physical assistance (e.g., for everyday tasks such as cooking and negotiating spaces at airports) and a lot of rest. For example, I adjusted to being a non-able mom to a young boy through strategies such as spending time together while watching comedy shows in response to dealing with symptoms such as pain, swelling, and extreme fatigue. This allowed me to rest and conserve energy but still be a mom – just a different one.

There were also times when my family members made me proud of their adjustment to having a disabled member of the family. My partner has always been ready to assist when needed and is a caring health professional. And my children learned to value people whose lives differ from the able norm. My son returned from public school one day and shared that he had been in the playground with friends when a DARTS (paratransit) vehicle went by. The other children laughed and made fun of the service only to have my son tell them to stop because, as he put it, "people like my mom need to use that." They stopped. I also think that having a non-able family member can help to instill respect for lives that are different in ways that help to promote compassion and care. Despite her busy work schedule, my daughter volunteers each weekend to bring her therapy dog to visit elderly people in a long-term care residence, which brings a lot of joy into their lives. And I think, in part, this relates to growing up with a non-able mom.

Of course, the consequences of being disabled and subject to ableism and intersecting forms of oppression are not always positive or happy ones, even in spaces of family and community life. For a number of years our son played ice hockey, and my partner and I loved to watch his games. While most venues were at least somewhat accessible to me as a wheelchair/scooter user, I do recall one recently built arena that was being used for a tournament and where I could only access the stands if I sat in a segregated seating section a long distance from the regular (able) arena seats. I remember sitting there completely alone, feeling isolated and wondering why on earth someone would design accessible seating in this way. Was it that the possibility of someone like myself coming to watch a game was an afterthought in terms of design or perhaps merely a way of keeping able patrons from being inconvenienced by wheelchair or scooter users (e.g., by needing more time to manoeuvre mobility aids)? I don't know what the answer is to this question, but I do know the message that such experiences in a space of family/community life send – that those who are different from the able norm will only be included on ableist terms – and that, in this case, this was segregation. If we think back to the civil rights movement this is not all that different from being forced to sit at the back of the bus because you were not white skinned. Despite some advances in creating more physically accessible environments, at least in affluent nations, there are still too many buildings with barriers such as stairs, automatic door openers that do not function, and lack of signage for persons with visual impairments that exclude and marginalize persons with impairments and illnesses.

Ableism, it seems, is everywhere and nowhere in our societies: everywhere in that if you are disabled, it seems impossible to escape this regime of privilege and power but, simultaneously, nowhere in the sense that so many people who are able remain oblivious to it.

As a visibly disabled woman I have been subjected to intrusive questioning by and even hostility from strangers while navigating places such as city streets and shops. On one occasion, a man I did not know shouted at me from across the street saying, "Get off the damn drugs!" The hostile tone of his voice was upsetting and made me feel that I was unwelcome on the street. I have had complete strangers ask me questions such as "What's wrong with your legs anyway?" and even "Who is your doctor?" For at least some people, encountering a visibly disabled woman invites hostile and intrusive questioning that they would not engage in with apparently able children, women, and men. This can be seen as another way of exercising power over those who differ from the ableist norm – pressuring them to be "out" in public spaces in ways that would not be asked of their able counterparts.

In any prolonged conflict there is what military personnel refer to as "collateral damage." By this they mean injury or damage inflicted on persons and/or places that were not intended targets. But while we at least sometimes recognize this damage in situations of armed conflict, it is too often ignored with respect to phenomena such as struggling for accommodation and freedom from discrimination in the workplace. In this section I share autoethnographic reflections, from my current vantage point as a senior feminist/social geographer and disability studies scholar, on the aftermath of these struggles in my life.

I met recently with a staff person in employee relations at my university. The impetus for our meeting was an enquiry received by the director of my academic unit about my adjusted teaching load. This raised alarm bells with myself and the director regarding whether I would have to fight yet again to keep accommodations, which were working, in place. The staff person and I talked about the history of my struggle and the accommodations currently in place (the major one being an adjusted undergraduate teaching load). He had taken the time to read some of what I had written about those struggles and, as our discussion ended, summed up the consequences by saying, "So you were set up to fail." While I had not thought about my experiences in quite that way before, in retrospect I agreed that he was right.

I share this encounter because it helps to illustrate some of the hidden trauma associated with ableism in the academy. As soon as the director informed me of the enquiry and expressed concern, I was angry, preoccupied, and on the defensive – worried that I would have to refight the battle for accommodation. I had difficulty concentrating on my work, which is another aspect of being "set up to fail" in an ableist academy. While I have not had a formal diagnosis, my hunch is that this is a form of post-traumatic stress disorder (PTSD) and is some of the hidden damage that is done when one is trying to contest an exclusionary and toxic academic environment. Once the threat of a new fight was gone, I reverted from my "warrior mode" back to my usual relatively calm self. However, experiencing that sudden shift back to the self I was during my many years of struggling for accommodation reminded me of how deeply distressing, disabling, and damaging life in the academy can be. Like collateral damage in war, wounds do not necessarily heal and losses may never be recovered. One example of the latter in my own case is the loss of a decade and a half of my academic career when I could have been doing more of the teaching and research I love but could not because I was too embattled not just with my own accommodation case but also with doing advocacy work on disability at our campus and in the community. The finality of all this hit home when the employee relations staff person asked if there was anything he could do for me. When I asked if he could give me all those years back, he just sadly and silently shook his head.

I want to be clear that I do not regret being active on issues of disability, gender, and class. While there are many losses involved in my own and other cases, there are also ways in which such adversity makes us better and stronger. For example, my own experiences in the ableist academy prompted me to engage in research about the challenges that other less privileged disabled people face. I thought that if I was having such difficulties in and outside the academy then these must pale in comparison to what most disabled people endure. And, unfortunately, I was right. I believe that being disabled also allows for a deeper appreciation of how ableism works in our societies and spaces of everyday life and that this is something we need to learn to value regarding disabled people. This is not to say that able-bodied researchers do not contribute important ideas about disability and ableism; rather, it is just to say that when it comes to understanding what it is like to be disabled and to live with the inescapable ableism in our societies, including disabled people in places such as academia provides a rich resource for learning and promoting social change.

I now want to return to the notions of loss and oppression in the ableist academy. I believe that we are still a long way from fully appreciating just how disabling and ableist the academy remains. I will address this through general observations and specific reference to my own experiences.

As discussed earlier in this chapter, academia remains stubbornly ableist with "success" largely being measured in quantitative terms that translate into the more you produce, the more you get paid and the more likely you are to be regarded as successful. There is insufficient recognition of the quality of one's academic work and a lack of recognition that the career path of a disabled academic is of necessity going to be different from that of an able-bodied academic due to aspects of impairment such as pain and mobility limitations and circumstances such as having to struggle for accommodation. My view is that such ableist ways of gauging success are outdated, oppressive, and, in the future, will be regarded as such. But, for now, I want to reflect a little further on the damage that ableist measures of success in the academy do. And for this, I return to my personal experiences.

From my early days as an assistant professor, my career and its success were questioned. First, as mentioned above, there was professional harassment from the chair of and the unknown peers in my department. Then, with becoming a visibly impaired and disabled female professor, came severe devaluation of my presence and career. I was told that the latter was over and the former unwanted. A salary frozen for a decade and consisting of one half in disability benefits was a response to such dismissive messages from an able vantage point. And while university administrators failed to enshrine salary discrimination into the new accommodation plan of 2003, the ableist academy had another way of enforcing an ableist academic order – the career progress and merit (CPM) system. In this system, professors compete with others in their home academic unit, mainly in terms of productivity, for above par rankings that translate into higher salary increments. In the fifteen years since the 2003 accommodation plan was finally put in place, I have only once received a "par" rating; all the rest have been "below par." This translates into low salary increments relative to those my able peers received over fourteen years and, in addition to a frozen salary during the 1990s, puts me in a position with a serious disadvantage in terms of salary. It is difficult, in such a context, not to internalize the notion that disabled professors are necessarily less "successful," and this indeed is some of the more insidious damage done by ableism in the academy.

I was reminded of this in a recent meeting with a staff person dealing with equity issues. I had asked her to speak with the director of my academic unit as to whether my accommodated workload was being considered in assessing my performance through the CPM system. He assured her that, at least under his watch, it was. But what struck me most about our discussion was how she summarized my immediate peers' view of my performance and career: "You are doing okay, sort of average but nothing stellar." It is important to note that she made this statement with the best of intentions (i.e., to help). But, even so, the ableism that pervades our world came through. One can only wonder whether such ableist and devaluing language would have been used if we in academia had learned to accept that disabled faculty's career trajectories will be different from that of their able peers and to value the fact that they make different but very important contributions to academic life, such as mentoring and teaching about the disabling and ableist world in which we live.

"Senior Moments": Reflections on Living in the Shadows of Ableness

Becoming a senior disability studies scholar has encouraged me to further reflect on the harms that ableness does in our lives within the academy and our societies. I am increasingly struck, for example, at how our encounters with ableness are manifest in all stages of our lives, albeit in different ways. Although we often associate advanced age with impairment and disability, it is important to remember that younger people are also vulnerable to the oppression of ableness – for instance, when they use cosmetic surgery in response to what they perceive as embodied "flaws."

In this sense, we all live our lives in the shadows of ableness. By this I mean that the perpetual struggle to approximate an able ideal helps to eclipse disabled people's accomplishments and their contributions to building a better world. Further, constantly being judged against this ever more elusive ideal, by ourselves and others, inflicts psychological and emotional "battle wounds" that are extremely difficult to heal. In my own case (see chapters 3 and 4), those wounds have included difficulties maintaining self-esteem, heightened pain, anxiety, and mental distress. And then there are the emotional costs of trying to resist being constructed as someone who has "failed" in academia and society. As I maintain in this book, however, there are fundamental ways in which the academy and society have failed me and other disabled people – for instance, by failing to recognize differences in career trajectories in academic and other workplaces and engaging in performance evaluations that directly compare able-bodied and disabled people's productivity.

I am also increasingly conscious that the human costs of ableness are too steep for us to continue to pay. People often recount "never feeling good enough" (e.g., for their job, attracting a partner, or accumulating wealth). This is a symptom of an able capitalist order that has struck at the core of our very way of being in the world. And the harm and suffering this causes is incalculable.

As a senior scholar, I am becoming more conscious, paradoxically, that although we are subjects of ableness throughout our lives, it is "old age" that tends to be most closely associated with failing to meet able ideals of embodied life. We fear aging not only because it is a prelude to our eventual death but also because of the losses it often entails for older people, such as diminished income, constructions of seniors as unproductive "burdens," and devaluation of their embod-iment and accomplishments. While there are some signs that images of old age are changing, for example, government advertisements pro-moting an active old age, we are still a long way away from making sure that older people are included and supported in our communities and societies as, for instance, appalling conditions in many nursing homes attest.

Conclusions: Imagining Life in a Non-ableist Academy and World

In conclusion, I want to briefly reflect on how we can build on the kinds of experiences I've shared to imagine what a non-ableist academy and world might be like.

First, we remain a long way away from a non-ableist academy and world. It is also clear that neo-liberal capitalism is implicated in maintaining this able order through punitive measures, such as slashing supports for disabled citizens and maintaining workplace cultures in which only quintessentially "able" performances are val-ued and rewarded. After all, neo-liberal capitalism is about extract-ing as much surplus value from workers as possible and imposing austerity measures on vulnerable citizens while allowing a small rich elite to amass even greater fortunes. This makes it difficult, if not im-possible, for most people and especially vulnerable groups such as disabled people to survive let alone thrive. The barriers to inclusion in academia and the CPM system that I've described in this chapter are examples of overt and more subtle ways of enforcing ableist priv-ilege and power. They also speak to how difficult it is to challenge this enforcement.

Second, we need to be imaginative and courageous in finding ways to do things differently in and outside the able academy. We have, for example, the technological means to make events such as conferences more inclusive of disabled academics through online presentations. Still, as my own experiences reveal, such presentations are devalued in comparison to in-person ones. Similarly, we have the knowledge and capacities to learn to value more positively the different career trajecto-ries and contributions of disabled scholars. We need to find the courage to say that to be different isn't necessarily to be less valuable and that when diversity promotes a more inclusive academy, this is something to celebrate and further encourage. We also need more progressive hiring practices to ensure that the diversity in our societies, including with respect to ability, is represented among those who teach and do research in the academy. We, our children, and future generations all have stakes in making this a reality and thus helping to ensure more just processes of knowledge creation and sharing.

Finally, we urgently need to take on the challenge of contesting disabling conditions of life in the world we share and the ableist order that perpetuates these conditions. At a time when social supports for disabled and other vulnerable groups are being cut back, when climate change is increasingly threatening human lives and health, and corrupt and despotic governments are on the rise, we badly need a change of course towards building societies that share wealth rather than hoard it, and societies that ensure the majority of people, including disabled people in the Global South, can lead more enabling and empowering lives.

4 Mapping Bipolar Worlds: Lived Geographies of "Madness" in Autoethnographic and Autobiographical Accounts

In this chapter, I draw on personal accounts of living with bipolar disorder or manic depression and what they reveal about how people contend with altered experiences of self, others, and ways of being in place that bipolar "madness" entails. Following an overview of the existing literature and the gaps in our understanding of living with bipolar, I share some of my autoethnographic reflections on life with this mental illness. This is followed by a thematic analysis of what eleven autobiographies reveal about experiences of the self-in-place and in relation to others and how these reflect the dynamics of ableist societies.

Bipolar disorder is a mental illness involving periods of extreme mental agitation with or without euphoria (i.e., mania or hypomania) and depression. Affecting 5 per cent of the world's population, it is the sixth leading cause of disability (Proudfoot et al., 2009). However, little is known about living with this illness. Proudfoot et al. (2009) observed: "Whilst there is an extensive research literature on patients' experiences of mental illness more broadly, there is a shortage of research on living with bipolar disorder" (p. 121).

There is also a need for more geographic research on, for example, how living with bipolar can alter encounters with others and the interpretation of attributes of place. When an earlier version of this chapter was published as an article in the journal *Health and Place* (in 2011), a search of this journal revealed only one article concerned with bipolar, and that article focused on inclusion and activity spaces rather than on how "madness" was lived in and through place. Even today, the same search in this journal revealed only an additional five articles. It is encouraging, however, that these studies are beginning to address negotiations of "madness" in place – for example, considering how young people with serious mental illness experience hospital wards in terms of emotions and interactions with others (Reavey et al., 2017). Analyses of how "madness" is lived are important – they provide a "window" into different psycho-emotional ways of being in place. They can render bipolar worlds more familiar from an "insider" perspective and, in doing so, can illuminate challenges of contending with this illness, including living with unmet needs. These analyses can unsettle stigmatizing conceptions of bipolar negotiations of place as exclusively "irrational." They can also allow us to explore lived identities of "madness" and the ways in which these may be contested or "rescripted."

This chapter addresses these gaps by considering the altered ways of being in place that bipolar "madness" entails and how narrative sense is made of these gaps. Conceptually, I build on Cosgrove's (2000) approach to psycho-emotional distress and geographic insights about being "mad" in place. Empirically, I draw on my own autoethnographic reflections on living with bipolar and on autobiographies of persons with this illness. It is worth noting that men's experiences are considered – something relatively neglected in the mental health literature (Ridge et al., 2011).

I argue that "mad" ways of being in place are paradoxically (dis)-embodied, sometimes discursively resistant and not straightforwardly "irrational." To place this discussion in a scholarly context, I provide an overview of the clinical literature on bipolar. This is followed by a discussion of how some scholars, including geographers, are trying to reframe experiences of bipolar and "madness" more generally as shaped by sociocultural relations and conditions. Next, I provide some brief autoethnographic reflections on my personal experiences of living with bipolar. This is followed by a discussion of aspects of life with bipolar that are illuminated through thematic analysis of eleven autobiographies. In conclusion, I discuss key findings and avenues for future research.

The Clinical Gaze: Living with Bipolar Disorder

Most clinical literature on bipolar adopts a biomedical gaze: exploring its etiology in chemical imbalances in the brain (Hallam et al., 2006), pharmacological treatments, symptoms such as disassociation (disruption in integrated functions of consciousness, memory, identity, or perception of the environment) (Oedegaard et al., 2008), and predictors of self-harm (Rihmer & Kiss, 2002). There is also work trying to pinpoint where in the brain this illness originates (generally thought to be the hippocampus). A comprehensive review of this literature is beyond the scope of this chapter. Instead, I focus on how clinical literature has advanced our understanding of life with bipolar and aspects of this that have been insufficiently explored.

One area of enquiry has been studies that relate symptom severity/ type to emotional/behavioural tendencies. Swann et al. (2007) found manic symptoms during bipolar episodes were associated with increased impulsivity and risk-taking (e.g., reckless spending). Cusi et al. (2010) showed that as symptom severity increases and psycho-social functioning diminishes, patients' capacities for cognitive empathy (taking another's perspective) decrease but distress at others' negative experiences escalates. These studies illuminated how characteristics of the illness experience relate to actions and psycho/emotional/social (dis)engagement. They revealed less about what it means and how it feels to contend with such facets of bipolar.

How living with bipolar influences self-identity is also being explored. Inder et al. (2008) identified challenges patients experience: confusion over who they really are, difficulty distinguishing self and illness, contradiction (e.g., having multiple mood-defined personalities), doubt about self-worth, and struggling for self-acceptance. This study showed how living with bipolar can hinder developing a cohesive and positively valued self but did not address how bipolar can also facilitate this and shape meanings assigned to being "mad" in place.

Studies of being diagnosed with bipolar have also emerged. Proudfoot et al. (2009) showed that patients grapple with diagnosis: in terms of ambivalence about taking medications, being terrified of or enjoying manic symptoms and feeling disbelief, shock, denial, anger, or relief at having a diagnosis to help make sense of their experiences. Inder et al. (2010) illuminated why patients ambivalently accept diagnosis due to frustration (e.g., having multiple diagnoses and being treated as "just another patient") and questioning their diagnosis/need for medication particularly when moods are stable. These studies demonstrated challenges in dealing with diagnosis but revealed less about discursive meanings assigned to being diagnosed and valuing/devaluing the "mad" self in place. Such insights are needed to grasp how bipolar identities are forged and contested.

Quality of life (QOL) with bipolar is also being examined. Michalak et al. (2006) noted that quantitative studies indicate diminished QOL but there is little qualitative research describing patients' experiences of how bipolar affects QOL. Drawing on interviews, they found that factors enhancing QOL included having a daily routine to help order a "chaotic" life or schedules flexible enough to accommodate changing mental states and social support. QOL diminished with difficulties maintaining independence (e.g., from family), experiencing stigma and needing to be careful disclosing their diagnosis. This study revealed how patterns/circumstances of everyday life impinge upon QOL. But close attention to how psycho-emotional, embodied experiences of illness and related ways of being in place shape well-being is missing.

Reframing Bipolar Socioculturally and Conceptualizing Being "Mad" in Place

Some studies have challenged biomedical constructions of life with bipolar by reframing this as in some ways being socially advantageous, not exclusively "irrational," and shaped by sociocultural processes of meaning-making. These help to revalue bipolar negotiations of place and put meanings assigned to this in a sociocultural context.

Jamison's (1994) *Touched with Fire* discussed the creative edge or advantage experienced by famous people apparently living with bipolar. As a psychiatrist with the illness, she has helped to popularize notions that bipolar is not only a liability. The biographical narratives discussed later in this chapter also sometimes reframe life with bipolar by stressing socially valued traits.

Martin (2007) reframed life with bipolar and attitudes towards it in America by putting these in a sociocultural context. She challenged conceptions that the lives of those with bipolar are inherently irrational/disordered – showing that persons also demonstrate rationality and keen insight. She probed popular fascination/affinity with, but also fear of, bipolar. In today's global capitalist order, bipolar traits such as strong drives for productivity are highly valued. But in the popular imaginary, this condition's "excesses" are also feared. For example, market fluctuations are often characterized as phases of mania and depression with opportunities for profit-making but also losses (e.g., due to "irrational" risk-taking).

Social geographies of "madness" have also sought to shed light on social and cultural forces shaping experiences of mental illness. The 1990s saw calls for greater attention to experiences of "madness" as involving an inner psycho-emotional world and negotiations of exterior places of everyday life (R.E. Butler & Parr, 1999; Parr, 1999; Parr & Philo, 1995; Philo & Wolch, 2001).

Parr and Philo (1995) showed that inner landscapes of "madness" were informed by specificities of place – such as asylums' proximity to villages. Parr (1999) argued that "delusional geographies" involved altered ways of being in place, including "unwilling" occupation of a distinctive world (i.e., visual, auditory, physical). Emotions such as compulsion were important facets of embodied experiences of delusional worlds as were finding safe spaces to perform "madness." This work provided valuable insights into how madness is lived. However, as discussed later in this chapter, some autobiographical narratives revealed that these worlds are not always engaged with "unwillingly." This work was arguably problematic in neglecting differences in being in place associated with particular mental illnesses. As a result, delusional worlds appeared experientially more homogenous than they probably are.

Efforts to illuminate the sociospatial dynamics of psycho-emotional distress continue. Knowles (2001), for example, showed how narratives about being schizophrenic drew upon realities of being in place and a delusional world of "imaginary" voices and visions. Personal stories about "mad" ways of being in place have a before and after structure as the onset of "madness" disrupts a former sane life and sense of self. These stories are shaped by discourses in the social environment (e.g., biomedical discourses of schizophrenia as a loss of an authentic self). Knowles's work showed how those with mental illness draw upon both delusional and non-delusional worlds in making narrative sense of their lives. Conceptually, I build on this by recognizing that this straddling of worlds can occur simultaneously – reflecting the fluidity with which the "real" and "delusional" are traversed by the "mad" self-in-place.

Geographic studies are enhancing our understanding of how meanings are assigned to places where "madness" is lived. Parr et al. (2004) demonstrated how familiarity among people can make rural communities threatening places for those with mental illness who, for example, risk anonymity when accessing clinical services. Tucker (2010) showed that the home can be a place of safety and refuge but also isolation for those with mental illness. While valuable, this work has been less effective in showing how meanings assigned to places of "madness" are not simply retrospective but involve prospective readings of anticipated events. This chapter addresses this aspect of lived experiences of bipolar.

Other studies revealed fluid boundaries between the embodied self and environment that psycho-emotional distress involves. Smith and Davidson (2006) showed how boundaries between self, body, and environment blur in experiences of phobia when feared objects (e.g., spiders) are felt to transgress bodily boundaries thus eliciting visceral reactions (e.g., feeling one's skin crawl). Davidson (2007) demonstrated permeable boundaries in autism as women cope with overwhelming sensory bombardment through calming practices such as rocking. In the autoethnographic and autobiographical sections that follow, I consider the paradoxical nature of this fluidity with respect to living with bipolar: experiencing place in ways that are complexly (dis)embodied.

The sections to follow also build on insights that lived geographies of "madness" sometimes involve "rescripting" the self and contesting stigma associated with "mad" ways of being in place. Davidson (2007) showed how women with autism seek out spaces, such as the internet, where they can validate a non-neurotypical identity. Parr (2008) considered how negotiating spaces of community gardening contributes to more and less validating or comfortable ways of being "mad" in place. The remaining sections in this chapter consider how persons with bipolar sometimes engage in "rescripting" "mad" ways of being in place. They are also informed by Cosgrove's (2000) feminist social constructionist/phenomenological approach to psycho-emotional distress. This perspective emphasizes how power, femininity, masculinity, identities, and psychopathology are produced through discursive practices. Subjective accounts are not straightforward revelations of a "true" self but produced/policed through social discourses.

Phenomenologically, building on R.E. Butler (1994), it situates agency in the "possibilities of resignification opened up by discourse" such that "we derive agency from the very power regimes that constitute us" (as cited in Cosgrove, 2000, p. 261). Thus, narratives of "madness" may fully or partially invoke and/or contest dominant representations of mental illness.

In the sections that follow, I recognize that "mad" ways of being in place are not unproblematically homogeneous but reflect diversity in illness experiences and how real and delusional worlds are negotiated or straddled. Meanings assigned to places of "madness" are not simply retrospective but also involve prospective readings. Experiences of "madness" are intensely embodied in some ways and yet may be paradoxically (dis)embodied – blurring boundaries of self, body, and environment. "Mad" ways of being in place involve (re)constructing identities; sometimes positively revaluing the "mad" self-in-place.

Living with Bipolar: Autoethnographic Notes

In this and the following section of this chapter, I want to reflect upon what we can learn from narratives about bipolar ways of negotiating the self, others, and place. I focus in this section on my own autoethnographic experiences of living with this illness.

As with many health conditions, bipolar disorder or manic depression is a stress-sensitive illness. Apart from my mother's unexpected death, all the other stresses that have negatively affected my mental health have been related to toxic work situations. The first episode was related to the harassment and overwork I experienced as a new assistant professor (described in chapter 3). In only two years, this had resulted in my hospitalization in the city's oldest mental hospital. It was complete with barred entrances and exits and included so-called "isolation rooms," and this made the psychic and physical experience of being there even more distressing. When in a paranoid moment I requested that the attending physician show me his identity card before treating me and then I said out loud that he was a student, he said this was unacceptable. I suppose he saw this comment as undermining his authority, but it was also a means of disciplining how I performed "madness." I was locked in a padded isolation room with only a mattress on the floor. I was forcibly injected with a tranquilizing medicine and had a terrifying night of hallucinations. At the time, I feared for my life and believed that at least some of my colleagues wanted me out of what was then the Department of Geography. The latter was certainly true and one way in which mental ill health is not straightforwardly "irrational." A day or two later I was transferred to the university hospital psychiatric ward where, even though I was still fearful for my personal safety, I recovered relatively quickly and without the bars that signalled confinement and made the old psychiatric facility such a distressing place to be.

This was in the late 1980s. Contemporary facilities are often less overtly carceral but still home to an array of disciplinary practices such as supervision of medications and behaviours. An example of this from these experiences of bipolar was how, when I tried to articulate that I was being harassed in my workplace and that this was the cause of my mental ill health, none of the nursing staff would believe me – but what they regarded as false was all too distressingly real to me. My colleagues in the Department of Geography also dismissed my concerns. And it made me feel even more vulnerable not to be believed. Perhaps I wasn't the only one straddling delusional and real worlds.

As noted earlier in this chapter, some scholars have noted that the boundaries between self, body, and environment/place can become more fluid in experiences of mental ill health. In my case, this was reflected in projecting my fear into all places of everyday life so that my mind and body literally hummed with anxiety. One example of this, following my discharge from hospital, was recurrent panic attacks – these were most pronounced, not surprisingly, at work but occurred elsewhere as well. And it took several years for this "fight or flight" response to ease – a reminder that the emotional scars associated with experiences of mental ill health can be long-lasting and difficult to overcome.

Although I was not hospitalized at the time, my mental health also deteriorated in the mid-1990s while trying to direct the women's studies program. I had accepted the offer of this position because I believed in the program and our students. However, it turned into a gruelling struggle to keep the program going, when influential people in the administration wanted to see it gone. It also happened that my geography PhD student from the United States at the time was found by a committee member to have plagiarized her thesis and this too added to my stress. The result was that I slipped into depression and remained depressed for two long years.

My symptoms included intense anxiety, especially at work but also in non-work situations, and difficulty concentrating and doing everyday tasks such as using my computer. The latter was both real and imagined – real in that I had had recurrent computer problems since taking on the role of director and needed our computer technician to fix problems, and imagined in that, in retrospect, I was no doubt making more errors due to concentration and anxiety issues. So this was a form of "straddling" real and delusional worlds. I was also very concerned about the future of the program I was trying to direct and with good reason. I felt utterly exhausted by this and so stressed by my student's plagiarism that I had to ask a colleague to fill in for me at the student's academic dishonesty hearing. I still went to the proceedings but let my colleague lead the discussion. Another PhD student of mine came along for support. The student who plagiarized was suspended from the university, and another colleague and I resigned our respective roles on the supervisory committee because of this academic dishonesty.

When I was too depressed to continue with directing the women's studies program, another colleague stepped in to do so, but it was eventually closed ostensibly due to small enrolment and small classes – some of the very attributes our students loved! But as a number of colleagues noted at the time, it was the so-called "bums in seats" approach to profit-making and managing finances. As one of a number of women who had helped to establish the program, I was also aware that there was some anti-feminist sentiment at work throughout the program's days, including, for example, male professors in the Council Chambers claiming the program would "ruin the university," professors and staff receiving hate mail, and graffiti appearing on the outside of the building and inside on surfaces such as hallways and doors. This was serious enough to warrant alarm devices for professors and staff at one point. Surely enough to help "drive one mad" – literally in my case. Fortunately, colleagues were able to develop an excellent graduate studies program in feminist research, which remains in place today and which should be a source of great pride for the university community. And I have enjoyed supervising the creative research projects developed by this program's students.

It so happens that I had another relapse (mainly depressive in nature) while writing this chapter. This relapse reminded me of how terrifying this illness can be and how the stresses associated with academic (and other) workplaces can help trigger relapse. With respect to the latter, I continued to work despite a spinal fracture that caused acute pain for about two months. This was because I wanted to contribute to important tasks such as hiring new faculty and a faculty retreat to discuss directions for my home academic unit in the future. The retreat was held in an inaccessible location for a scooter user and so I arranged for a rental scooter to be delivered to the retreat venue. Unfortunately, the driver bringing the scooter to the retreat was in an accident and so I did not receive the scooter until 2 p.m. (the retreat began at 8:30 a.m.). I was able to use an old manual wheelchair until the scooter arrived, but this did not allow me to navigate the venue independently. While this accident was likely a random event, I began to worry that the inaccessible location was a deliberate strategy to limit my ability to participate in the retreat. This worry seemed to make sense given the long history of my marginalization in my unit. And it is in this way that people sometimes talk about experiences of "madness" as being themselves only amplified.

As my mind feverishly worked to try to make sense of such a scenario, other deep-seated fears were triggered and exaggerated. I feared for my life and the well-being of my family, for example, particularly at the hands of my estranged sister who has always wished me ill. Such fears were not straightforwardly "irrational," then, but were amplified so as to be all-consuming and exhausting or, as I said to my psychiatrist, "I feel as if World War Three is going on in my head." While I was able to recuperate on additional medications, I tired easily, my concentration was still impaired, and I remained more depressed than usual. In some ways, recuperating from mental illness can be much harder than recovering from a physical illness or injury. Perhaps this is, in part, because it is easier to assess one's progress in recovering from physical conditions, such as pain, than it is with "invisible" illnesses such as bipolar.

This episode also blurred the boundaries between self, body, and place/environment. I was consumed with intense fear for myself and my family and that made me fearful of being in places such as a passport office, a plane, work, and so on. The only salient features of place that seemed to register with me were whether strangers were about who might harm me and whether someone was there that I trusted to keep me "safer." In this sense I experienced a kind of placelessness where fear was pervasive.

Even today, it is painful to recount how toxic my working environments have been for so long and how this has contributed to my mental as well as physical illnesses. So I won't delve further into my experiences here. Except to say that today, I am fortunate to be in a more collegial and inclusive academic unit. This partly reflects increased concern with diversity and inclusion on campus. But what can we learn from the experiences I've shared? A first lesson, perhaps, is that the actions of medical professionals can sometimes make the experience of being in a mental health facility even more distressing. I realize that some encounters are distressing to health professionals as well, and this is understandable. But the more we know about and can relate to the experiences of those with mental health challenges, the more we can gain empathy and respect for them and the richer our encounters will be. My first experience of hospitalization occurred in 1989, and there have undoubtedly been significant advances in treatment since then, including designing hospital psychiatric units to be more therapeutic in their appearance and operation.

A second lesson is clearly that stresses associated with working in academia cause real harm to real people. Some of these stresses may be inadvertent but others are deliberate, as in the lack of support among administrators for an accommodation agreement and for developing an accommodation plan for me for so many years. Academic environments are stressful enough to begin with, deliberately making them more so for individuals deemed "out of line" with the "status quo" is unacceptable. This form of assault, which is largely invisible, harms professors, staff, and students and threatens the viability of universities themselves in the sense of missing out on the rich insights into our world that diversity makes possible.

Exploring Other Voices of Bipolar "Madness": Autobiographical Accounts

Thematic analysis of eleven autobiographies was used to explore negotiating bipolar in society and space. Books were selected through amazon.com, chapters.com, and ebooks.com. Five authors were male and six were female.

Accounts were manually coded using a constant comparative method. Coding proceeded by identifying key substantive incidents in narratives (e.g., first becoming "mad"). These were compared across narratives to identify similarities and differences. An iterative process of theoretically categorizing and revisiting coded data was used to help illuminate how madness was lived in place. This has similarities to

Glasserian grounded theory (Walker & Myrick, 2006) in that it allows categories to emerge from data but differs since thematic categories such as invoking dominant discourses about being mad were partially derived from the theoretical framework and were pertinent to conceptual insights of the existing literature (e.g., experiencing a "fractured" or "cohesive" sense of self).

These accounts offer rich insights into life with bipolar but, as Davidson (2007) noted with respect to autobiographies of autism, they are not necessarily representative (e.g., most likely representing "higher functioning" persons). Using autobiographical accounts also risks producing more "distanced" analyses of the illness experience than methods involving researcher/participant interaction. Nonetheless, they provide one "window" into lived geographies of bipolar.

"Fast Forward": Experiences of Mania

Mania refers to episodes of extreme mental and emotional agitation. In mania, a person's psycho-emotional life and negotiations of society and space are "revved up" or put on "fast forward." This section explores key facets of these lived experiences.

1. Blurred Boundaries of Self, Body, and Environment: "Flying"
and Assigning Meaning to the (Dis)embodied Self in Place

A heightened pace of engaging with the world and sense of movement through space are important facets of mania. Autobiographical accounts often use metaphors, such as "flying" or being on a "runaway train," to characterize being manic. Mariant (2006) described the onset of mania during a business trip in the following way:

> [I]t was at the beginning of the trip that I began to notice unusual changes. For example, my creativity and speed of thought increased noticeably while I was flying. It was as if I was flying metaphorically as well as literally ... It was an election year, and I was thinking about becoming the president of the United States – not something I normally thought about. I also thought about making a movie, along with other fantastic plans. Such thoughts are termed "grandiose," and they fulfill one of the diagnostic criteria for mania. (p. 14)

Accentuated creativity and rapid thought foster a sense that the mind/self is metaphorically "in flight" in making plans and negotiating places such as work environments – aspects of mania that involve (dis)embodied experiences of a frenzied mind and yet are anchored in being in place (e.g., flying on a business trip). His narrative invoked biomedical discourses of illness, abnormality, and diagnostic criteria to make retrospective sense of rapid thinking, movement, and fantastic plans as part of mania. This constructs the "mad self in place" as pathological, giving medical authority to the illness account. But there are elements of narrative "resistance" as well. He stressed how creativity contributed to "productive" ideas in his subsequent business meeting, thus partially revaluing manic ways of being in place.

Hornbacher (2008) described mania as a (dis)embodied state of frenetically chasing racing thoughts and yet experiencing visceral embodied sensations (e.g., shortness of breath, rapid pulse). Boundaries between body and environment seem to dissolve:

> The moods are steady sky-high. My mind is racing ahead and I chase it, writing as fast as I can, failing heart stuttering, body disappearing. I can do anything. Nothing can stop me. I'm a flurry of motion, sitting on the floor of my bedroom arms flying, shuffling papers into piles, brain racing, reading snippets of writing ... Making rapid little red pen marks on the pages, cutting and pasting, short of breath, pulse pounding, I am back in my element, where I can do a thousand things at once, fueled by the rabid energy triggered by the booze, no food, no sleep. (p. 38)

Her narrative underscored the mental and physical intensity, and yet paradoxically (dis)embodied nature, of being manic in place – blurring boundaries of self and environment. Like the previous author, Hornbacher invoked dominant discourses of being manic in place as "abnormal," as phrases such as "rabid energy" indicate. Her narrative also reflected rescripting of "mad" ways of being in place in terms of abilities to achieve, as statements such as "I can do anything" indicate.

Estrada (2009) wrote of experiencing mania in college, and described it as her normal mind "slipping away" and a frenzied mind taking over:

> I was under a lot of stress. I had so much anxiety. I developed insomnia and then became very manic. Then suddenly, my mind began to slip away. It felt like a whirlwind that would not stop spinning me. It took me over, spit me out into the world of the madly insane. (p. 22)

Estrada's "before and after" narrative revealed the intensity of mania and complexly (dis)embodied ways of being in place. How being taken over by an ill mind feels like a disorientating "whirlwind" that spins her rapidly blurring boundaries of self/body/environment and shifting her into a "mad" world. She made retrospective sense of this by invoking biomedical discourses, juxtaposing sanity and insanity.

Coping with distress at facets of mania such as frenzied thought sometimes included repetitive behaviours aimed at inducing calmness (cf. Davidson, 2007, on autism). Hornbacher (2008) wrote about repeatedly cutting herself to calm racing thoughts:

> I am numb. I am in the bathroom of my apartment ... twenty years old drunk and out of my mind. I am cutting patterns in my arm, a leaf and a snake ... I study my handiwork. Blood runs down my arm wrapping around my wrists and dripping off my fingers onto the dirty white tile floor. I have been cutting for months. It stills the racing thoughts, relieves the pressure of madness that has been crushing my mind, vise-like, for nearly my entire life. (pp. 1–2)

Her account illustrated how efforts to calm the self are in some ways (dis)embodied (e.g., being "numb" to bodily pain, feelings) and yet involve bodily awareness (e.g., bleeding). In this and some other accounts, alcohol and/or drugs were also used in efforts to moderate frenzied "highs" (and depressive "lows") of negotiating bipolar worlds.

2. Straddling Real and Delusional Worlds and Assigning Meaning to Places Where Madness Is Lived

While dominant ableist discourses oppose "madness" and "sanity" as discrete abnormal versus normal ways of being in place, accounts of mania reveal that those living with bipolar in some ways "straddle" (i.e., have an experiential foot in) both "real" and "delusional" worlds. One is recognizing some of the behavioural implications of being "mad." Bishop (2009) wrote wistfully of mania:

> [A]t my wildest, I somehow had an inkling I was over the edge ... In one of my tapes I mentioned the beauty of the paintings of artist Tom Curry and reminded myself that though I was crazy now and adored his work, I should check it out when I wasn't crazy and see how I felt about it. Those little glimmers [i.e., of rationality]. (p. 80)

Narratives of mania also reflected a "straddling" of real and delusional worlds by drawing upon both "real" and "imaginary" attributes of places and people and using logic/reason to conceptualize, in compelling ways, what is going on in the world and its implications. These experiences of the self-in-place are emotionally intense – including dread, terror, and euphoria – and elicit actions such as fleeing places and people. Mariant (2006) described feeling dread for his son's well-being and was compelled to find an item while manic:

> Outside I felt a sense of dread, for some reason, and became afraid for my son's life. I also felt that I needed to get a copy of a Barbie game on CD for my daughter. She had been talking about the game, and it seemed linked to saving my son's life somehow. (p. 17)

Later, he described how, while psychotic, he believed city schools were concentration camps for children and that he must demonstrate allegiance to Hitler:

> After my wife picked me up from the hospital, things began to get really weird. As soon as I got home, I ... drove off to have breakfast with my friend Frank. After breakfast, I drove around ... and noticed concentration camps all around ... What I actually saw were schools ... getting major re-modelling done. There were chain link fences all around to provide a safety barrier. That is why I thought they had the appearance of a concentration camp ... While I was driving and while eating at the restaurant, I felt Hitler's atrocities were still going on ... kids were being marched off to concentration camps to be executed ... At the restaurant, I thought the hamburger patties were made from human flesh. I thought that I had to eat the human flesh burgers to show my allegiance to Hitler. (p. 53)

As these passages illustrate, delusional experiences often make logical sense as part of a wider conception of what is going on in places of daily life, how one should relate to others, and observations about environments. Because delusional ideas are parts in a wider conceptual logic, manic ways of being in place feel compellingly real. Mariant's narrative assigns meanings to places: retrospectively reading school landscapes as places of execution, and prospectively reading the restaurant as a place where, without demonstrating loyalty, one is in danger.

Psycho-emotionally intense ways of being in place were also associated with "mixed" illness states (combining symptoms of mania and depression). Diven (2009), a physician, wrote:

> It is hard to articulate what those first few months of mania were like ... Anxiety and constant worry were predominant. Nothing is more mind occupying than anxiety. I felt as though at any moment something terrible was going to happen ... Sleep came with difficulty despite exhausting days and nights [as an intern]. I felt as if I was in a whirligig of fast thinking and

activity with a depressed and anxious mood. If I wasn't crazy already I was going to go crazy feeling like this. (pp. 54–5)

Noting that experiences of mania include not just chaotic thinking but also goal-oriented activity (challenging notions that being "mad" is wholly irrational), Diven goes on to relate fear triggered by paranoia:

Bipolar disorder may cause chaotic thinking in some ways but very goal-oriented focus in other ways. I was very focused on my work. My speeding mind easily welcomed the excitement of racing around town lights flashing [to accident scenes]. A new feature popped into my mind: paranoia ... It really just showed up in my consciousness without warning ... I was frightened at times of people I had to work with. At first the focus was the police department. Later ... my own paramedics. Ultimately I was afraid of Nazis and Chinese agents it felt real and frightening ... [At the scene of a collision] ... What did occur to me was that the police were probably preparing to capture me. I feared the supervisor's small talk was intended to keep me standing around until they had enough officers to take me down. I hurried away. (pp. 74–5)

Here again is an experiential straddling of "real" and "delusional" worlds: Diven recognized being engaged in small talk and that a few officers were in place but contextualized such "facts" through paranoia about potentially being "taken down." Here, again assigning meanings to places where "madness" is lived is not only exclusively retrospective but also prospective: anticipating future events and their implications for being in place. It shows how altered senses of the self-in-place as vulnerable elicits emotions and actions that "make sense": fearing and fleeing dangerous people and places.

3. Being "Broken" or "Whole": Fractured and Cohesive Sense of Self and Ways of Relating to Others in Place

As noted in some clinical literature on bipolar, and flagged in Parr (1999), a fractured sense of self and ways of relating to others are facets of being "mad." This is often, but not always, borne out by narratives about manic ways of being in place. Melcher (2008) wrote of becoming "wobbly" in mania: shifting away from a "real" self that is taken over by a less acceptable masculine "alter-ego" and relating to others in ways empowering to him but troubling to them:

This form of "wobbly," most noticeable to Sandra [his partner] was a sort of "Dr. Jekell and Mr. Hyde" phenomenon where, with no apparent provocation, my personality would shift from "Mr. Nice Guy" to "Mr. Jerk" instantly. Utterly confusing, disturbing and frightening for her – strangely empowering, self-determining and freeing for me. Only ... it wasn't ... the healthy me! It was an alter-ego in which the illness ... ruled my psyche – my thoughts, feelings and actions at these times of stress and lost-ness. (p. 15)

This altered self is discursively constructed as "not me" but paradoxically understood as bound up with the psycho-emotional life of the "real me." Thus, the "mad" self is and isn't "me" also signalling a fractured sense of self. He continued:

I call this extreme manic alter-ego "Sham" ... Sham is the unleased expression of years of unexpressed feelings: anger, fear, disappointment frustration, humiliation self-pity – all rolled up into a rolling lead ball, spinning down and crushing anything in his way! He's a bundle of the worst parts of me exposed and in action. (p. 15)

Later, he described this alter-ego as the "broken" part of me (p. 19). Characterizing Sham as a destructive version of the "real self" constructed his "manic" self and ways of relating to others as defective. This illustrates how assigning meanings to mania involves compartmentalizing and valuing/devaluing aspects of the self-in-place.

If the "me" ruled by a manic alter-ego draws on the "worst" parts of "me," then non-mad ways of negotiating place must draw on the "best." There is an evaluative "fracturing" of the self that affirms some and rejects other facets of the "real me." This narrative again illustrates how dominant discourses about "madness" become woven into personal accounts of being ill (e.g., constructing the self-in-mania as less authentic).

Cheney (2007), who described "rapid cycling" (shifts between mania and depression), also wrote of a fractured self and how it felt to assume multiple different mood-defined personalities: "From the moment I awoke I had been a quivering mass of volatility: up, down, irate, flirtatious, contentious, giddy, seductive, paranoid. I'd assumed half a dozen different personalities between daybreak and dusk. No wonder I was so tired" (p. 81). At play in her account are tensions between more and less acceptable feminine ways of being in place and relating to others (e.g., being more acceptably flirtatious versus being irate). Her narrative highlights how a fractured sense of self feels and is experienced bodily as a tired, volatile, quivering "mass." The latter also perhaps suggests a blurring of experiential boundaries between the self, body, and environment – as the volatile self is experienced bodily but also as a mass of energy/activity pulsating outward.

As in other accounts that signalled a fractured self, Weiland (2009) wrote about, in the wake of manic behaviours, feeling foolish and defined by a "spoiled psychotic self." Her narrative constructs the manic self as defective and expresses a longing to escape illness:

> Every morning when I woke up I hoped that it was over ... Some days the mania was so intense that trying to relax felt like running a marathon and being told to stop and meditate right at the finish line. My heart raced, my head was spinning yet lifting my eyes or even a pencil required full concentration. I was supposed to sleep as much as possible ... but because my head was in overdrive, I couldn't completely shut down. Thinking about the headlines, about frightening my children, about putting our lives in reverse ... made me feel like a fool. I'd worked hard to better my life. Now I was just seen as a spoiled psycho. (pp. 163–4)

Although I cannot provide a detailed exploration in this chapter, feelings of shame, regret, and foolishness following manic behaviours (e.g., erratic dangerous driving) were common themes. Revealing another self who related to the world in frenetic ways left people feeling exposed and vulnerable.

Narratives of mania did not always reveal a fractured sense of self and discursively constructed "abnormal" ways of relating to others in place. Mania also sometimes involved a more cohesive or "whole" sense of self as connecting in valued ways with others, enjoying "epiphanies" and even feeling at one with the heavens. Such ways of being in place elicited powerful feelings of euphoria and well-being. Bishop (2009) wrote of great joy at feeling the mania resolution of life's problems:

> I arrived very happily at the town dock, looking merrily at the stars, talking to myself a mile a minute. There seemed to be so much to say. Never had I been filled with so much complete joy. It was so wonderful having extensive conversations with my long-deceased parents, whom I had adored. Most of my life's enduring problems were getting important recognition from the heavens and were resolved. Even the worst one, my childlessness, became a non-issue that night. Of course, no one could have two birth mothers – so my god-daughters and my niece and nephew just lucked out and landed with a second, just as important mother: me. What a relief to reach that conclusion. How sad that my epiphany on this score and others disappeared with the mania. (p. 21)

In this narrative, senses of reconnection with loved ones and heavenly resolution of problems, notably feelings of loss at not fulfilling the gendered role of motherhood, elicit intense joy and relief. However, it is inflected with sadness and loss as joyful insights and unconditional self-acceptance dissipate along with manic ways of being in place. She wrote longingly of these aspects of her "madness":

> How I wish those experiences could in fact be controlled, that my "epiphanies" were still a part of me – some of course were nonsense, however some were glorious. What I especially miss of … mania is the extraordinary loss of the Watcher – the critical observer who has analysed and harshly judged my every move, my every word, since my early twenties. During my manic states, the Watcher disappeared and with it my self-consciousness. I was treated to the awesome marriage of my intellect and … soul. I rejoiced, I danced and sang, I was fiercely intelligent and clever in social interactions. (p. 22)

Manic ways of being-in-place can thus involve a strong sense of empowerment, an ability to relate to others, and a blurring of boundaries of self/environment, eliciting feelings of being at one with the heavens and the world. Such valued experiences challenge characterizations of delusional ways of being in place as involving necessarily "unwilling occupation" of a "mad" world (cf. Parr, 1999).

Religiosity in mania contributes to a more cohesive "whole" self that experientially merges with a religious figure or entity. This heightens senses of meaning and purpose in relating to people and places. Melcher (2008) wrote:

> I realize now that bipolar creates a whole elaborate covey of false relationships and situations – often backed up by some religious goal for me to live out the life of a biblical character … When in high mania I have often risen into this kaleidoscope world of time-warps and world-saving. (p. 29)

Here, manic ways of being "mad" in place involve being caught up in a "kaleidoscope" that blurs the past and present. The present self merges with a past religious figure creating a grand sense of purpose (e.g., he wrote about travelling to a cemetery in order to save its "lost souls").

Other narratives of religiosity do not convey a temporally fluid "mad" world. Instead, the "mad" self merges with enduring religious entities (e.g., God). Hornbacher (2008) related:

> On this summer morning I experience that defining (psychotic delusional) break. I go from bipolar II to bipolar I just like that. A doctor might put it this way: I go from sick to really, really sick. For the average Joe, I go from having an illness "just like diabetes" to being flat-out crazy. But

I, cheerfully mad as a hatter, am entirely unaware that something has snapped and will never be put back together. Here we are, it's Tuesday, and we are quite mad. Not mad as in moody. Mad as under the impression that I am God. (p. 71)

Her account of being "mad" invokes biomedical discourses and popular juxtapositions of "crazy" and "sane" ways of being in place. By invoking the "Mad Hatter" she retrospectively captures what being in a delusional world means: although something is broken in herself, she does not realize this.

Estrada (2009) described being in a delusional religiously meaningful world when taken to hospital: "I thought the devil was in the fireplace and he could see me ... I went in and just stood before the fireplace and said 'I hate you Satan!' I thought I was the chosen one. God chose me to save the world" (p. 24). Here, a cohesive sense of self is forged in terms of being God's "chosen one." Her narrative illustrates how this self straddles "real" and delusional worlds: for example, noting hospital features (e.g., fireplace) but making sense of them in religiously meaningful ways. It shows how manic ways of being in place, such as expressing hatred towards Satan, make sense as a part of a wider conception of what is happening and its implications for how one should negotiate people and places. It illustrates how persons retrospectively construct thoughts and actions that seem logical at the time as an expression of a "crazy" self. She also assigns prospective meaning to people and places while manic: after encountering "Jesus," she constructs the ward as a place where they can join forces against Satan.

4. Rescripting "Mad" Identities and Manic Negotiations of Places of Everyday Life

Narratives of mania primarily draw upon dominant discourses of "madness" as being defective and abnormal. Less commonly, authors struggled to rescript manic identities and negotiations of places as socially valuable. This provides insights into how identity formation involves contesting meanings assigned to being manic.

Bishop (2009) emphasized how a cohesive sense of self in mania involved an awesome synergy between intellect and soul. This was associated with euphoria, joyful performances of the self-in-place, and feeling especially adept at engaging with others: "During my manic states ... I was treated to the awesome marriage of my intellect and my soul. I rejoiced. I danced and sang, I was fiercely intelligent and clever in social interactions" (p. 22). Manic ways of being in place can thus elicit a strong sense of empowerment, ability to relate to others, and celebratory responses to feeling "whole." These allow Bishop to rescript her manic identity, and value qualities such as social skills. This affirms for her a sense of belonging in places of everyday life.

Rescripting manic ways of being in place were sometimes evident in narratives predominantly framed by dominant conceptions of madness and which signalled a fragmented self. Elements of discursive rescripting of identity were evident in Diven's (2009) narrative about manic negotiations of place: insisting that thinking and action are not exclusively "chaotic" but can lead to socially valued outcomes (e.g., being focused and productive in the workplace). Blurring the boundaries between "madness" and sanity, revalues such facets of being manic.

Such revaluing of manic negotiations of place challenge claims that bipolar straightforwardly diminishes the quality of life. They shed light on how forging more positive "mad" identities draws on a wider cultural narrative affirming traits such as creativity and drive. Mariant (2006) wrote of how the "tremendous sense of creativity" associated with mania often "fools the person with bipolar into thinking all is well" (p. 16) – both positively reclaiming madness and more negatively flagging why this can make bipolar worlds seductive. He challenged stigmatizing conceptions of "madness" reclaiming a positive identity and sense of belonging among the "great":

> [B]ipolar disorder has actually taken on some very positive meanings for me ... I find it fascinating how many leaders, artists – all expressive and intelligent people – are believed to have suffered from depression or bipolar disorder ... If you suffer from bipolar disorder or severe depression like me, we are in good company. We are counted among the greatest of the great people to have ever walked the earth. (p. 16)

Feelings of intense joy, creativity, brilliance, productivity, and social aptitude, and a cultural milieu partially affirming such manic ways of being in place, help explain the ambivalence of some persons living with bipolar towards the illness experience, including diagnosis and taking medications to help moderate manic "highs."

"When It All Comes Crashing Down": Being "Trapped" and "Weighed Down" in the Dark World of Depression

Autobiographical narratives focused primarily on experiences of mania – reflecting, perhaps, greater opportunities for injecting drama into accounts of being "mad" and rescripting mad identities in more positive ways. Nonetheless, the slide into deep clinical depression is as much a part of lived experiences of bipolar. For some, such as Weiland (2009), depression predominates as a state of being in a bipolar world. She wrote of seeking herself-story in celebrities' autobiographies, focused primarily on mania, and finding "parts of my story" but "I haven't found myself" (p. 173).

In the autobiographical accounts reviewed, depression has been portrayed as a psycho-emotional spiral downward from the heights of mania into the depths of hopelessness and despair. Hornbacher (2008) likened her flights and falls of mood to riding a rollercoaster of thoughts and intense emotions:

> The past few years have seen me in ever-increasing flights and falls of mood, my mind first lit up with flashes of color, currents of electric insight, sudden elation, and then flooded with black and bloody thoughts that throw me face down onto my living room floor, a swelling despair pressing outwards from the center of my chest … I have ridden these moods since I was a child, the clatter of the roller coaster roaring in my ears while I clung to the sides of my little car. (pp. 1–2)

Her narrative suggests that lived geographies of depression, like those of mania, are complexly (dis)embodied and involve at least some blurring of self and environment. She responds bodily to the "flooding" of her mind by dark thoughts (i.e., hurling herself to the floor) and yet experiences despair as embodied and (dis)embodied: as swelling inside herself but also pressing outward from her chest blurring boundaries of self and environment.

Stringham (2007) also contrasted the "sparkling state" of being manic with the dark depths of depression: "My moods were like the ocean tides. Either high or low or moving toward high or low. Never stable. Always pulsing and throbbing. The sparkler state of mania. Or the dark flannel of depression" (p. 84). For her, feeling the sadness of depression involves psycho-emotional responses such as shame and fear. Moreover, boundaries between self and environment blur as sadness alters experiences of place: diminishing light in her room and sometimes eclipsing light altogether: "I am ashamed and afraid for me, too. That when I feel sad the light shrinks in my room. That when I am really sad, even the sun goes dark" (p. 84). Her narrative reveals that her feelings of shame and fear about depressive ways of being in place are linked to a sense of being up against a threatening world and, reflecting her fragmented sense of self, beliefs that she had become evil even monstrous. She continued to explain: "The world is an enemy some days, often many days on end. That I am evil. That my reflection in the mirror is the devil ... That the monster I found hiding in the shower today is me" (p. 84). Her narrative of depression, like many of mania, draws on long-standing negative conceptions of being "mad" as monstrously "abnormal."

Negotiating depressive ways of being in place included feeling trapped and immobilized emotionally, mentally, and physically. Estrada (2009) wrote of being suicidal and depressed in terms of a sense of being trapped and weighed down by chains:

> I often sit in my room and think of ways to successfully kill myself. Manic depression is deadly. Even today at the age of twenty-eight, I still find myself suicidal. I feel like there is no way out ... When I explain depression to people I always say that it is a plague and it feels like pulling chains. You pull and you pull but it seems impossible to move forward ... It's not that you don't want to be happy, it is that you are incapable of being happy. (p. 16)

Her narrative draws on biomedical discourses of disease and "contagion" and illustrates how depressive ways of being involve embodied sensations such as pulling chains.

Cheney (2007) wrote of being suicidal but not acting on it because "suicide requires movement and depression weighs a thousand tons" (p. 166). Similarly, Diven (2009) wrote:

> My depression came on quickly and with no warning symptoms. It was as if I just woke up depressed one morning. I felt as if I had a hangover from all of life. I wasn't suffering from just the blues; I was suffering in a dark, deep place that had close walls. In very little time I ached with despair and feelings of hopelessness and helplessness. I soon felt suicidal. (p. 11)

As these accounts have illustrated, experiencing bipolar depression involves complex (dis)embodied ways of being in place – for instance, feelings of despair that seem to darken the world around you.

On Being Bipolar in an Ableist World

I now want to focus on what we can learn from these bipolar narratives about living in an ableist world.

One insight, particularly from the autobiographical narratives considered above, is what a strong grip biomedical and devaluing discourses have on how we imagine and characterize life with bipolar: as being crazy rather than sane, defective, and abnormal, and even monstrous and evil. And yet, paradoxically, there are traits associated with mania that are positively valued in our able neo-liberal capitalist order – heightened creativity and productivity are prized for contributing to innovation, profit-making, and the accumulation of wealth (at least for the elite minority).

We are all, able and disabled people, under pressure to outperform each other so as to advance in the workplace in terms of remuneration and authority. If we acquiesce to such pressures, we risk creating more toxic work environments and damaging the health and well-being of ourselves and our families. These are manifestations of what Berlant (2007) calls "slow death": where we are ground down by difficult living conditions and the excessive performance demands of an able neo-liberal capitalist order.

Another insight, this time from my autoethnographic notes, is how important it is to pay attention to invisible assaults, for example, in the workplace, that help to trigger mania and depression. It is a fact of life that, in our neo-liberal and frenzied world, we are constantly forced into competition with each other and, along with that, experience having our abilities compared to those of others. This is as true for able people as it is for disabled people and ultimately results in many people feeling "never good enough." This is an appalling situation and one that fails to recognize that people have diverse gifts, as well as perhaps limitations, and that the point is to make the most of the former and accommodate the latter so as to build a more inclusive and enabling society.

These narratives also help to demonstrate that characterizations of life with bipolar as exclusively "irrational" and "crazy" fail to acknowledge that madness entails straddling both a real and delusional world. So, for example, one can register characteristics of places such as how barbed wire encircling schools under renovation makes them look like concentration camps but then make the conceptual leap that this means the need to demonstrate allegiance to Hitler. In our ableist societies, there is the frankly intolerant notion that to be "mad" is to have no claim to rational thought, which heightens the stigma and shame that people with illnesses such as bipolar experience.

Ableist pressures to conform to compulsory able-bodiedness and mindedness are also manifest in the complex (dis)embodied experiences of mania and depression reported. While manic, for example, I felt as though my body and mind hummed with anxiety and projected related fears that I and my family were in danger onto various places of everyday life. Others characterized their experiences of mania in terms of figuratively "flying" or being trapped on a runaway rollercoaster where frenzied (as opposed to able, calm, and rational) thoughts and activities hurled them into a strange, "mad" world. Transgressing pressures to conform to compulsory able-bodiedness and mindedness also occurred during depression: feeling "weighed down" as if by chains and so unable to act and experiencing a darker more hopeless world.

The workings of an able, neo-liberal capitalist order are also evident in the link between toxic work and other environments and mania and depression. These are not exclusively "individual" phenomena but relational such as, in my case, being marginalized at work for so long. I have argued that we need to learn to see such "invisible assaults" as causing real harm to real people. Such assaults are efforts to construct people as "less able" and "less valuable" in our societies and spaces of everyday life. And they are likely to become more frequent in our increasingly precarious work environments and societies more generally.

In the next chapter, my emphasis shifts from experiences of mental ill-health and from the personal and interpersonal scales to a consideration of forces perpetuating ableness, impairment, and disability in a neo-liberal global capitalist order. This helps to provide a basis for the following two chapters that deal with the pursuit of able embodiment and the limits of human rights law as a basis of securing sociospatial justice for disabled people. Still, I would encourage readers to keep the personal in mind, especially in regard to thinking about how the forces I discuss are impacting their own lives and the lives of diverse able and disabled people.

5 What's Neo-liberal Global Capitalism Got to Do with It? Ableness and the Production of Impairment and Disability

In this chapter I explore the linkages between the workings of neo-liberal global capitalism and the perpetuation of ableism or ableness, impairment, and disability. My primary focus, as indicated earlier in the book, is on our contemporary global capitalist order and some of the dangerous forces it has unleashed. These not only produce, in a geographically uneven fashion, impairment and disability but also help to perpetuate the ableism on which, in part, the global order is based. But before addressing this topic, I first want to reflect on lessons that can be learned from volume 1 of Marx's *Capital* (1867/2013) about the kinds of forces we need to be attuned to when exploring these connections today.

Lessons from Marx's *Capital*

What can we (still) learn about the dynamics of capitalism from volume 1 of Marx's *Capital*? One of the wonderful things (figuratively if not literally) that we can learn from this work is the centrality of violence against labouring bodies and minds that is at the heart of capitalism – in the violent extraction of surplus value from those workers, who, without access to the means of production (capital in its various forms), must sell their labour-power to the capitalist class. The capitalist then pays the workers their wages for their labour and appropriates the difference between those wages and what the capitalist can realize in terms of profit by selling the commodity or service produced. This difference in value is what Marx termed "surplus value." The forms this violence take have in some ways changed since Marx's industrial capitalist era when, for instance, dangerous working conditions in factories resulted in high rates of physical impairment (e.g., losing limbs as a result of working with factory machinery and doing so under pressure from factory managers to work harder and faster) and exposure to toxic substances in lines of work such as pottery making resulting in illness and premature death (Marx 1867/2013).

Marx, quite rightly, had scathing things to say about the brutal extraction of surplus value from working bodies. Commenting on the labour process and working day, he noted that capital is dead labour that, in order to continue to accumulate wealth, "vampire-like" sucks the life out of labouring bodies. Although he did not use the terms impairment or disability, Marx was clearly concerned with the violence and harm the working day and organization of the labour process was inflicting on working-class bodies and minds. Harvey (2018), in a similar manner, compared capital to a mosquito that sucks out the life from so many working bodies. Marx (1867/2013) had the following to say about the harm capitalism was doing to the health of workers and how indifferent it was to that harm:

> In every stockjobbing [a speculative transaction on the stock market] swindle everyone knows that some time or other the [economic] crash must come, but everyone hopes it may fall on the head of his neighbour, after he himself has caught the shower of gold and placed it in safety. Après mois, le déluge! is the watchword of every capitalist and of every capitalist nation. Hence Capital is reckless of the health or length of life of the labourer, unless under compulsion from society. To the outcry as to the physical and mental degradation, the premature death, the torture of over-work, it answers: Ought these to trouble us since they increase our profits? (vol. 1, chap. 5)

While capitalism and social conditions of its existence have changed in many ways since Marx published *Capital* in 1867, it remains the case that central to this mode of production is the antagonistic, exploitative relation between capital and labour – a relation that can destroy people's capacities to work and live in able ways and that consigns the disabled worker to the ranks of what Marx termed the "lumpenproletariat" or labour that is surplus to the needs of capital in specific times and places. In the discussion that follows, I aim to keep these facets of capitalism in view when considering the dynamic inter-relations between capitalism, ableism, disability, and impairment. In the next section, I discuss some of the linkages we can make between neo-liberal global capitalism and the production of impairment and disability. This is followed by a discussion of connections we can make between this order and the perpetuation of ableism.

Making Connections I: Neo-liberal Global Capitalism and the Production of Impairment and Disability

Today, the wearing out of labouring bodies and minds, what Berlant (2007) termed "slow death" (as discussed in chapter 2), is still manifest in the health conditions discussed by Marx (1867/2013) but increasingly these take the form of chronic health conditions such as cancer, heart disease and stroke, chronic respiratory illnesses, and accidents of various kinds. According to the World Health Organization (WHO; 2020), in 2018 there were an estimated 9.6 million deaths due to cancer worldwide. Seventy per cent of those occurred in low- and middle-income countries – countries that tend to have little or no cancer treatment available. In 2018, only 30 per cent of low-income countries reported having treatment programs, compared to 90 per cent reported in high-income countries.

Such disparities in cancer deaths and in access to treatment are among the legacies of a neo-colonial global capitalist order that continues to disadvantage less affluent nations and their peoples. It is important to acknowledge that the wearing down of bodies and minds that occurs in countries of the Global South is especially difficult to address for a number of reasons, including the greater prevalence of infectious diseases and of places (e.g., hospitals) with conditions that encourage the transmission of these diseases to others. WHO (2020) estimates that infectious diseases in low- and middle-income countries are responsible for 30 per cent of all cancer deaths. Other factors limiting access to care and ultimately survival include the outmigration of health professionals and facilities (e.g., mental health centres) that have too few resources to address the needs of people dealing with mental impairment and illnesses (Chouinard, 2019).

But the basic point is still that this unevenly developed capitalist order injures, impairs, and kills workers labouring but, so long as they can be replaced (by other workers or technological innovations) and commodities continue to be consumed, this need not hamper the continued accumulation of wealth by a small elite. This is the harsh truth of our contemporary capitalist order: that the killing and maiming of workers still lies at the heart of its dynamics and that members of the capitalist elite are, for the most part, unapologetic about this reality – a reality that reflects a political economy in which profits are too often valued over the lives of people.

So, what are some of the similarities and differences in the dynamics of capitalist oppression that we need to keep in mind when exploring connections between compulsory able-bodiedness and our contemporary global capitalist order? One, as Harvey (2014) and others have noted, is that the capitalist system has become global in scale (which is not to say that there aren't pockets of resistance). This means, among other things, that ideas, information, money, and investment circulate almost instantaneously. And that, thanks to technological "advances" such as mobile phones, tablets, computers, and of course the internet, many people can be tuned or embedded into these flows for more of the time. This can lead not only to demands on labourers becoming more intense but also to greater pressure (e.g., virtual) being put on them to consume an increasing array of commodities and services. This too helps to perpetuate our neo-liberal global capitalist order.

In such a global order, where transactions are speeded up, for instance in the finance sector and stock trade, giving in to such pressures as to be able-bodied and even hyper-able have arguably become increasingly compulsory. Nor are such pressures limited to these sectors. Recent reports on working for Amazon (a global online seller of a rapidly increasing array of commodities) suggest that workers face multiple pressures to work harder and more "ably" and that this has been associated with numerous physical injuries and impairments as well as mental health issues and even death. A recent letter, drawing on an investigation by thirteen members of the US Congress into working conditions in Amazon warehouses in the United States, made the following observations:

> Hundreds of stories shared with our offices paint a picture of desperation and a corporate employer with little regard for the health of its employees. We heard from an Amazon worker who described the warehouse as a "21st century sweatshop." Workers shared stories of high temperatures in some warehouses and an inability to take water or bathroom breaks for fear of retaliation: "I myself take medication so I will not have to use the restroom and drink little fluids also to help. Sometimes I think is this torture really worth it??"; "[being] afraid to drink water for fear of not hitting [his] rate"; "[having] no air conditioning when it's hot in the facility"; "[having] the air conditioning ... not working and you could easily pass out from the heat in Arizona going on 115 that week"; "please help stop the mandatory overtime and ten hours shift in a humid atmosphere." (Sanders & Omar, 2019a, p. 2)

Another worker was reported to have issued the following desperate plea: "Please send OSHA [Occupational Safety and Health Agency] to investigate these conditions that are affecting workers physically and mentally" (Sanders & Omar, 2019a, p. 2).

The letter goes on to note that despite frequent injuries and deaths, at least some of which are not reported, workers remain under intense and dangerous pressure to speed up the pace of their work and meet their "rates" or output. In such situations, compulsory ableness morphs into "hyper-ability," placing workers' lives and continued capacities to work at high risk (Sanders & Omar, 2019a; see also Sanders & Omar, 2019b). Corbett (2019) decried these conditions of work (and associated injuries and deaths) and the injustice of the fact that one of the wealthiest men in the world, Amazon's founder, Jeff Bezos, lives in luxury while Amazon's workers suffer, become impaired and disabled, and even die because of inhumane working conditions.

Inhumane and toxic work environments put those engaging in paid labour within them at risk of impairment, illness, and becoming too (dis)abled to continue to work. This was certainly the case for myself – stress exacerbated the pain associated with rheumatoid arthritis and, as I explained in chapter 4, triggered bouts of mania and depression. As noted in chapter 2, work is becoming especially stressful with the rise in precarious jobs. These are defined as irregular (or intermittent) part-time, contract, and/or freelance work with little or no benefits. Insecurity makes this form of employment especially stressful and thus potentially injurious to health and capacities to engage in paid labour. These jobs are not limited to low-paid service work but are found in other sectors as well (e.g., professional). A study by the Canadian Centre for Policy Alternatives found that 22 per cent of Canadian professionals are in precarious work (over one out of every five). The study also found that the majority of these professionals (60 per cent) are women (Hennessy & Tranjan, 2018, p. 5). It was reported that "precarious work cuts across all employment sectors, professional occupations, wage levels, ages and career stages" (p. 7). Given that women already face discrimination in employment (such as lower wages and fewer opportunities for promotion), the fact that they hold the largest proportion of precarious professional jobs is worrying in terms of compounding their vulnerability to work-related impairment, illness, and (dis)ability. There is also cause for concern regarding precarious work and its "spill over" effects at the global scale. On this, Quinlan and Bohle (2009) observed: "Since the late 1980s, a growing body of international research has found an association between precarious employment and a deterioration in wages and working conditions (including worse occupational health and safety outcomes)" (p. 1).

Neo-liberal austerity policies are causing poor working conditions, even in vital sectors such as health care in countries such as Canada and the United Kingdom. A recent report on the care of seniors in residential institutions in Ontario, Canada, noted that personal support workers (PSWs) have as little as five or six minutes to care for each resident and that they understandably think that they are being asked to provide levels of care that they have not been given the resources to achieve (Frketich, 2020). Sometimes care tasks such as baths are skipped in the struggle to deal with time constraints (para. 12). Wages are little more than minimum wage and add to the stress of trying to do an impossible job. This staffing shortage has prompted outcries against inadequate long-term care:

> The union and the Ontario Health Coalition (OHC) have released a report that alleges long-term care homes are short one to two PSWs on almost all shifts. It adds up to at least five to 10 PSWs short in a 24-hour period.
>
> "In every town, in virtually every long-term care home, on virtually every shift, long-term care homes are working short-staffed," it states.
>
> "There is a PSW crisis in long-term care in Ontario that is worse than we have ever seen," it warns. "The staffing shortages threaten care and safety for residents and staff alike."
>
> Unifor says the PSW shortages are so significant that funded long-term care beds in Thunder Bay have not opened. The union worries that the same will happen in Hamilton. (Frketich, 2020, paras. 5–8)

This is a situation in which older people are being "warehoused" in facilities without the resources needed to support them. As the preceding quote recognizes, it is not just care that is being undermined but also the safety of residents and staff. A 2017 case in St. Joseph's Villa in Dundas, Ontario, illustrates the grave consequences that can ensue when large institutions such as the Villa are underfunded and understaffed. The case involved an elderly man who resided in the Villa. He was sleeping in his room at night and, without any PSWs nearby, a violent man with dementia, also a resident, wandered into his room. He proceeded to punch the man so viciously that the resident was black and blue all over. He and his family were deeply traumatized. The man died of his injuries two months later (Anonymous, personal communication with author, April 2017). While no staff were reported injured, one can see that they are at high risk of being harmed as a result of understaffing and poor working conditions.

Other workers, deemed essential in pandemic and non-pandemic times, include migrant farm workers. These workers travel from countries such as Mexico and regions such as the Caribbean to assist in farm work and harvesting on farms across North America, sending money back to families in their home countries. There are an estimated 60,000 such migrants currently in Canada (Johnson, 2020). Wages are low and working and living conditions are poor, and migrant farm workers are paying the price in terms of their health and abilities. Conditions such as crowded bunk houses and insufficient sanitation and other protective measures have allowed outbreaks of COVID-19 to emerge. Government officials have denounced the conditions in which migrant workers live, but it is not yet clear what steps, if any, will be taken to improve their working and living conditions (Antonacci, 2020).

What Harvey (2018), among others, has termed "accumulation by dispossession" is also becoming more prevalent in our global neo-liberal capitalist order. By this he means that when opportunities to amass wealth through investing existing capital decline, then capitalists may reappropriate assets held by others to try to kick-start economic growth and further the accumulation of wealth. A good example from the 2007–9 recession is when US banks embarked on high-risk, speculative mortgage loans and then precipitated a housing crisis through foreclosing on those mortgages and seizing homes (Harvey, 2008).

In addition to workplaces that help to precipitate impairment and illnesses, it remains the case that many persons with physical and mental impairments and illnesses are disabled through exclusion from the workplace. An article published by the Canadian Centre for Policy Alternatives characterized this phenomenon in the following way:

> Canadians with disabilities face exceptionally high rates of unemployment. Over 400,000 disabled working-age Canadians are currently unemployed despite being willing and able to work. While Canada's unemployment rate is currently sitting at about 5.8%, the rate for disabled Canadians is much higher. Canadians with "mild" disabilities are most likely to find employment, and their unemployment rate is 35%. For those with "severe" disabilities, the rate jumps to 74%. Put another way, for every one person with a "severe" disability who finds work, three do not. (Raso, 2018, para. 5)

In the United States in 2018, the proportion of disabled people employed was 19.1 per cent compared to 65.9 per cent employment for those without a disability (US Bureau of Labor Statistics, 2019). As of April 2023, the employment rate for disabled people aged sixteen to sixty-four in the US showed some improvement at 38.3 per cent. However, this still paled in comparison to the 77.4 per cent employment rate enjoyed by able-bodied people in the same age range (US Bureau of Labor Statistics, 2023).

In countries of the Global South, such as Guyana, the employment rate remains low for persons with impairments and illnesses. In an interview with *Guyana Inc.* in 2016, Leon Walcott, chair of the Guyana Council of Oganisations for Persons with Disabilities, pointed out:

> The population census of 2002, deemed the last detailed census, noted that despite no job discriminations against disabled people in Guyana, they did not have the same opportunities to participate in the labour force. Of the 42,577 disabled persons (15 years and over), 22% (9,388) were in the labour force, that is, were economically active and were working or had the need for work. ("Disability in the Workplace," p. 26)

Labour-force participation/employment rates for disabled people in Guyana are significantly lower than those for all working age people. In 2017, there was a 54.5 per cent labour force participation rate (World Bank Group, 2020, Economic Transformation section, para. 1). While my own research on disabled people's lives in this country reveals that there is job discrimination against disabled people in Guyana by, for example, paying disabled people less than would be paid to an able worker, the finding that 22 per cent were in the labour force and that this included people seeking employment suggests that the unemployment rate is underestimated.

The antagonistic relation between capital and labour that produces impairment and disability is part of the everyday workings of what scholars such as Klein (2008) have referred to as "disaster capitalism." This takes multiple forms, including increasingly difficult and toxic workplaces, but also lucrative destruction of infrastructure such as hospitals (e.g., as in Palestine) and profitable rebuilding and the deployment of military personnel and equipment. Impairment and disability are also perpetuated through harsh disciplining and unsafe working conditions often, but not always, in less affluent parts of the world. The Rana Plaza garment factory collapse in Bangladesh in 2013, in which managers forced workers to continue to labour despite concerns that cracks in the wall meant the facility was extremely unsafe is a case in point (Fitch, 2014). The death toll from the collapse was over 1,100 while another 2,500 were injured and impaired. This was a situation in which labourers were viewed by those managing the factory as disposable and easily replaced. It was also a situation in which the factory was poorly constructed and unsafe. These practices point to the disregard for workers' lives intrinsic to capitalism in Marx's time and in our own. While there have been some improvements in working conditions in the textile factories since the collapse, only some factories have benefited from these. S. Butler and Begum (2023) estimated that two million workers are covered by the recently implemented Bangladesh Fire and

Safety Accord but another two million workers are not. The Accord is an agreement among various actors in the textile industry (e.g., managers, workers, and brand name clothing firms). But fires also threaten the safety of workers as do other working conditions such as sexual harassment. And, as is the case elsewhere, there are difficulties achieving justice for the workers killed and injured by the collapse. While the Rana Plaza owner remains in prison, reparations to those affected by the disaster have been held up in the courts (S. Butler & Begum, 2023). This is true with respect to other industrial accidents, such as the Union Carbide gas explosion that I discuss next.

The Union Carbide gas explosion in Bhopal, India, that took place thirty-five years ago is an example of an industrial accident that continues to take its toll in perpetuating impairment, ill-health, and suffering. An estimated 3,000 people died immediately from exposure to the poisonous gas. Women exposed to the gas have had significantly higher rates of still-births and miscarriages and those who have given birth often do so to babies with various impairments and illnesses. Diseases such as cancer and heart and lung disorders are also rampant among those exposed to the gas. One woman described what life is like for survivors:

> "It would be better if there was another gas leak which could kill us all and put us all out of this misery," said Omwati Yadav, 67, who can see the Union Carbide factory from the roof of her tiny one-room stone house, painted peppermint green with orange doors. Her body shaking with sobs, she cries out: "Thirty-five years we have suffered through this, please just let it end. This is not life, this is not death, we are in the terrible place in between." (Ellis-Petersen, 2019, para. 4)

Sadly, efforts to bring the corporations responsible to justice have failed. The Indian government has not acknowledged the scale of destruction caused and it, along with the United States government, has acted to protect corporate interests. Nor has the site been decontaminated despite offers from other countries to assist with this (Ellis-Petersen, 2019).

In the Union Carbide case, the worst industrial accident on record, we have a very stark reminder of the degree to which capitalist corporations and governments show disdain towards the life and health of workers and others affected by the explosion.

We can also point to examples of exposure to toxic contaminants in wealthier nations of the world. Gordon Plaza, now a Black neighbourhood in New Orleans, Louisiana, was a site used to dump industrial waste from the 1900s to the 1950s. The city then began to plan for government subsidized housing on the site. Those who bought the houses there in the 1980s and 1990s were never told that their homes and neighbourhood were built on a toxic site. When the Environmental Protection Agency began to test the site in the 1980s, it found 150 contaminants in the soil and forty-nine of these were known to cause cancer. Not surprisingly, the cancer rate is high in Gordon Plaza: "Earlier this year, a report by the Louisiana Tumor Registry found the census tract in which Gordon Plaza is located had the second highest cancer rate in Louisiana, at 745 cases per million people, compared to a state average of 489" (Zannolli, 2019, para. 17). Although residents have won two class action suits as well as a suit against the City of New Orleans and its mayor, they have not received compensation or been resettled. As one homeowner put it, "We are just waiting to die." An environmental activist who lives in Gordon Plaza put it somewhat differently: "We're in the fight for our lives" (Zannolli, 2019, paras. 40, 12).

Despite being in different parts of the world, the cases discussed above all speak to a callous disregard of human lives and well-being in our global able capitalist order. Gordon Plaza is a reminder that the production of impairment, disability, and ableness also marks some bodies as more or less important and disposable than others. In the cases discussed above, those bodies are devalued on the bases of race and class as well as disability. And this is done in the name of feeding insatiable appetites for profit-making in our able neo-liberal capitalist world.

Contemporary neo-liberal austerity policies, policies favouring unfettered free market capitalism, cutbacks in government support for already emaciated social programs, and increased tax cuts for the super-rich are also helping to fuel death and impairment. In our current able neo-liberal global capitalist world, we are seeing austerity measures that are rendering more citizens "disposable" and consigning them to lives of poverty and neglect or, at worst, death. In the wake of the COVID-19 pandemic, we are also being exposed to the reality of a struggling health care system. One consequence of this is a shortage of necessities such as ventilators. And at least at some hospitals, doctors are deciding who will live and who will die among those coming into the emergency wards with the virus.

In one recent case in the US, doctors decided, in consultation with the state, that a disabled Black man with a brain injury and paralysis was not enjoying "quality of life." This was even though he was alert and able to talk and joke with loved ones. Nonetheless, a "life not worth living" in the eyes of doctors and state officials was the rationale for taking him off the ventilator he needed and using it instead on a younger and non-disabled patient. The disabled patient died despite protests from his family about his care (Flanders, 2020). There are also reports from countries around the world of elderly people's lives being regarded as "disposable" in hospital settings and even in their own homes through domestic violence, which saw a dramatic increase during the pandemic (Summers, 2020). On the situation in hospitals, Summers (2020) reported:

> When Souzi Bondeko's grandfather started showing symptoms of Covid-19 and was struggling to breathe, she took him to a hospital in the Democratic Republic of the Congo's capital, Kinshasa, where he was put on a ventilator.
>   She dashed home to get some food and returned to be told by a member of staff that he had been taken off the machine as it was needed elsewhere.
>   "There were only three ventilators in the hospital and they were in great demand," she said. "Five minutes later, my grandfather died. Staff told me they had to give the ventilator to a younger man as it is their policy to prioritise younger patients." (paras. 1–3)

She continued:

> The 33-year-old market trader reported her grandfather's treatment to Anatole Bandu, the country representative for the charity HelpAgeInternational. He said: "Unfortunately this is not the first instance of older people dying as a result of ventilators being given to younger people. Sadly, older people are seen as dispensable in DRC." (para. 4)

Discrimination against older people has been on the rise since the pandemic, according to the organization. It has heard reports from many countries of older people left to die from COVID-19 as younger patients are prioritised – or simply being refused treatment amid fears they would infect others (Summers, 2020, para. 5).

Harsh cuts in government spending for groups such as disabled people and the elderly are, as I argued in chapter 2, forms of violent cleansing in societies where the potential contributions of such groups are ignored/dismissed. A particularly heinous example in Ontario, Canada, were efforts to cut spending for services provided to autistic children. The outcry as well as legal action against the Ontario government from parents over cutbacks was so intense that the government is apparently reconsidering this move (Canadian Press, 2019). Conveying the depths of human suffering precipitated by benefit cutbacks to vulnerable people requires that we all listen to their stories. In a recent example in the UK, Errol Graham, grappling with depression throughout his life, had his benefits cut off completely even though his mental state worsened if he ventured out of his flat. This situation, coupled with anxiety, prevented him from going out to work or on some days from even tolerating noise such as that from his washing machine. Without income or food, Errol died of starvation – when he was found dead in his flat by persons coming to evict him, he weighed only sixty-eight pounds (thirty kilograms) and was fifty-seven years old (Pring, 2023). Errol's story is one of a multitude worldwide that demonstrate how austerity kills, impairs, and disables.

It is also important to acknowledge that climate change, including global warming of the planet, contributes to illness, impairment, disability, and death. Famines and extreme weather events, rising sea levels, the acidification of oceans, the loss of species diversity, and increasingly fragile ecosystems (e.g., coral reefs) are making it more difficult for humans and other species to survive let alone thrive. It is as if the violence towards labouring bodies intrinsic to capitalism has morphed into something even more sinister and distressing – a global order increasingly prone to killing life as we know it in the name of profit and the accumulation of wealth for those who are part of a global elite. The United Nations Global Issues (n.d.) has summarized the climate crisis in the following way:

> Climate change is the defining issue of our time and we are at a defining moment. From shifting weather that threatens food production, to rising sea levels that increase the risk of catastrophic flooding, the impacts of climate change are global in scope and unprecedented in scale. Without drastic action today, adapting to these impacts in the future will be more difficult and costly. (para. 1)

Climate change impairs and kills – not only humans but also the many other animal and plant species we too often take for granted and which are part of our fragile ecosystems. To share just one example of this is that as sea ice in the Arctic melts, polar bears face increasing difficulty hunting for food. Their main prey, seals, rely on the ice to care for their offspring. The melting of ice in the Arctic thus endangers the bears' food supply. The bears, increasingly desperate for food, are going hungry and even venturing into human settlements to try to find food. The World Wildlife Fund (n.d.) has estimated that, if current trends continue, the polar bear population may decline 30 per cent by 2050. People living in remote settlements are also affected since they have lost the ice that makes hunting and travelling possible in the colder months of the year and are increasingly isolated as a result.

As Nordhaus (2013) convincingly argued, the best of climate science and economics has suggested that it is prudent to act now to limit CO_2 and other green-house gas emissions that encourage global warming. As he noted, though, it will take decades to see the benefits of policies such as "cap and trade" (capping carbon emissions and allowing companies to trade their carbon allowances to others) on limiting climate change. Moreover, many risks and uncertainties remain. We don't know much about "tipping points" at which change is likely to accelerate very quickly and in ways that will be difficult to predict. Nor is there an easy or inexpensive technological "fix." And, of course, influential people, including former US president Donald Trump and even some scientists, claim that climate change is a "hoax," which helps to dissuade members of the public from supporting efforts to combat global warming before it is too late (see Klein, 2015). Indeed, the intensity of the debate over climate change is such that climate scientists are reporting death threats and other forms of online abuse on social media such as Twitter (now known as X) (Valero, 2023).

Unchecked, climate change will almost certainly exacerbate levels of impairment and disability not just for humans but also for other animal and plant species. For example, along with famine and droughts, we will see impairment and disability as a result of malnutrition and shortages of clean water (the latter fuelling the spread of gastro-intestinal diseases); along with acidification of the oceans, we will see loss of marine life on which other species such as fish depend for food; and along with extreme weather events, such as the 2019–23 wildfire catastrophes in countries such as Australia, Canada, the United States, and Spain, we will see more people and animals killed and impaired and disabled.

To the best of our knowledge, then, climate change impairs and disables. And it arguably reflects an arrogant brand of ableness that presumes humans have the right to exploit all of the earth in the name of capitalist profit-making and consumption. As climate activists such as Greta Thunberg remind us, however, we will be failing future generations and the planet if we do not adequately respond to the disabling crisis unfolding before us.

Making Connections II: Neo-liberal Global Capitalism and Ableism

The production of impairment and disability is, of course, a reflection of how central ableism or ableness is to our lives and struggles. I have defined ableism as a regime of power through which quintessentially able forms of embodiment are valued and favoured over non-able ways of being. As noted in chapter 2, Wolbring (2008) has acknowledged that some abilities are prized more than others in our global capitalist order. Not surprisingly those at the top of the hierarchy reflect traits most highly valued in neo-liberal capitalism, such as individual competitiveness and the accumulation of wealth. There is thus a "pecking order" in ableist regimes of power in which abilities such as empathy are comparatively devalued.

In this section I want to further illustrate some of the ways in which neo-liberal global capitalism is linked to the perpetuation of ableness. A belief in the superiority of ableness and related relations of power, are, as some of my previous remarks in this chapter begin to suggest, at the heart of contemporary able neo-liberal capitalism and its contradictory tendencies. In terms of political economy and culture, we are heavily invested in approximating an able-bodied norm.

From everyday practices such as "hitting the gym," in our efforts to embody ableness (and health), use of plastic surgery and cosmetics as we seek to appear younger than we are, to practices such as refusing to use aids such as wheelchairs (even when these are needed), most of us strive to project as able an identity and place in the world as possible. Why do we do so? One reason is arguably continued beliefs in the perfectibility of humans, first through the eugenics movements of the twentieth century and now the genetic and medical advances and debates of the twenty-first century. The latter include the rise of phenomena such as "medical tourism" (Connell, 2006; C.M. Hall, 2011; M.D. Horowitz et al., 2007) whereby more affluent patients from wealthy countries travel abroad for surgeries that are, in their view, more timely and/or effective in making them "able" once again. One facet of this transnational movement of people seeking to extend ableness and health, the illegal organ trade, also reminds us that in terms of well-being, poorer "donors" are the ones who lose out in the trafficking of body parts. This will be discussed in greater detail in chapter 6.

I've also argued that the ableism most difficult to see/acknowledge is that which some of us experience while having an able body and mind. One example that I gave in chapter 3 was when my intellect was assessed and reassessed to see where I belonged for schooling purposes. This is one form of ranking and assessing children and a microcosm of our general practices of sorting humans out in comparison to one another. Who is more or less intelligent, who is more or less talented at activities such as sports or drama, who is the best "worker of the month," and who is most rich and successful? Although these comparisons are being made all the time, so long as we are able-bodied we remain relatively oblivious to them. This is, in part, why it is difficult to see ableist privilege for what it is: central to what we do but also a way of reinforcing a narrow conception of "normalcy" and constructing disabled bodies and minds as defective and inferior (Titchkosky & Michalko, 2009).

But why is ableness so deeply entrenched in the dynamics of our contemporary global capitalist order? One reason is that being an "able" labourer is perceived as being a more profitable and productive one whether or not this in fact is the case. Another is the increasingly widespread adoption of neo-liberal austerity measures. These have further eroded the social safety net established by governments after the Second World War and in doing so has emphasized instead that "success" (including well-being and health) is achieved through individual merit rather than political action. This has also shifted the onus for assistance for groups such as older people onto unpaid family members – this latter shift is seldom, however, acknowledged among politicians and policy makers (Cameron, 2020).

Ableness also is buttressed through the consumption of commodities that promise to enhance our attractiveness to others. This includes the consumption of clothes and accessories, makeup and other cosmetic products, perfumes, and services such as hair salons and spas. It also includes doing our best to approximate able ideals of the body and mind through our consumption practices (e.g., working to approximate the slender woman and muscular man and the intelligent, productive, or creative woman and man). Those who cannot afford to consume such commodities and services face the risk of being constructed as "lesser" and less desirable than able others. This is a harsh mode of able rejection.

Our consumption of commodities also has cultural capital in demonstrating our economic and social "success." The most obvious commodity in this regard is the home, which through its size, various features, amenities, and more or less desirable location, helps to highlight how successful one is. Another such commodity is vehicles – luxury cars with desirable design features are often seen as singling out those who have "made it" in neo-liberal global capitalism and those who have not.

However, it is important to keep in mind that these processes are unfolding at a time when the extreme accumulation of wealth by a privileged elite is accelerating and reaching historic heights. This is happening as most people's share of wealth has declined. In other words, "success" in contemporary capitalist societies is a very relative term. An article by Phan (2020) in *The Guardian* reported that more than 70 per cent of the world's people are experiencing a widening gap between their own economic assets and the wealth of the rich. Phan went on to say: "Social and economic disparities have soared even in countries such as Argentina, Brazil and Mexico, where inequality had been falling in recent decades" (Phan, 2020, para. 2).

Statistics on the global distribution of wealth help to illustrate the wealth gap at the global level:

According to the Credit Suisse Global Wealth Report, the world's richest 1 percent, those with more than $1 million, own 44 percent of the world's wealth. Their data also shows that adults with less than $10,000 in wealth make up 56.6 percent of the world's population but hold less than 2 percent of global wealth. (Inequality.org, n.d., Global wealth inequality section, para. 1)

Conclusions: From Impairment, Disability, and Ableness to Human Enhancement

This chapter has focused on key linkages between impairment, disability, and ableness in our contemporary global neo-liberal capitalist order. It has focused on forces such as capitalist exploitation of labour power, precarious work, dangerous working conditions, climate change, and war as causes of impairment and disability. This order is sustained and perpetuated through ableness (or ableism), which I have argued is a pervasive regime of power and oppression in our contemporary capitalist world. Like other privileges, ableness tends not to be noticed by those with able bodies and minds, making it difficult to contest.

In the next chapter, I consider how ableness finds expression in efforts to be more able than others (even beyond current able norms) as people jostle for position in places such as the workplace or school. As Brashear (2013) has shown, those who support this trans-humanism are prepared to consume products that enhance human memory, human speed, and provide more sophisticated prosthetics – all of which are seen as allowing human beings to outperform their peers and thus "succeed." This is, I argue, ableness run amok in the sense of waging war on any person who does not wish to and/or cannot afford to consume such enhancements. Or, in other words, it is a kind of contradictory process whereby ableness turns in on itself to generate new standards of ableness that do not even leave persons with able bodies and minds unscathed. This raises challenging issues. How far, for example, are we prepared to go in using science and technology to "fix" or perfect the human body and mind? Are pressures growing to be ever more able in the best interests of ourselves and our societies? In answering such questions, we need to keep in mind that escalating requirements to be "able" are fuelled by capitalist profit-making and the accumulation of wealth, for instance, through the pharmaceutical industry, procedures such as stem cell treatment, and the production of prosthetics and mobility aids. And that there are important class and geographic inequalities in who does and does not have access to such commodities and the greater "ability" they are seen to promise. Indeed, ability itself has become both increasingly elusive and commodified.

6 To the Ends of the Earth: Wealth, Poverty, and the Pursuit of Able Embodiment

This chapter considers another central facet of the able neo-liberal global capitalist order, namely, the ongoing commodification of human embodiment itself in the name of profiting from and accumulating wealth through what is sometimes termed "human enhancement" (Brashere, 2013; Clarke et al., 2016). I argue that the promotion and consumption of a large array of products and services is part of the pursuit of able embodiment and human enhancement. These products and services range from anti-aging creams, vitamins, and pharmaceutical drugs to fitness equipment and services such as gyms, spas, and hair salons. This promotion and consumption also includes services that are transnational in scale such as medical tourism and international trade in and transplantation of human organs. I use the example of the human organ trade to illustrate how this allows more affluent citizens of the world to purchase bodily enhancements at the expense of poorer people in nations of the Global South.

The chapter begins with a brief discussion of the debate about human enhancement. This is followed by discussion of forces contributing to the ongoing commodification of human embodiment in the able neo-liberal global capitalist order; using examples to illustrate some of the many forms this takes. This is followed by a discussion of the particularly troubling global trade in human organs for transplantation and some of the ethical and political dilemmas this raises regarding what the limits to this form of human enhancement should be. This is difficult terrain involving as it does more affluent, ill patients seeking to prolong their lives and well-being and sellers who are desperate to escape debt and poverty by selling an organ.

Debates about Human Enhancement

What is meant by human enhancement? While we lack a definition that everyone can agree with, it is generally understood to involve the use of drugs, medicine, and technological means to enhance the mental, emotional, and physical abilities of human beings (Clarke et al., 2016) sometimes beyond what is currently "normal" for the species. Clarke et al. (2016) summarized the debate in the following way:

> We can enhance some of our mental and physical capacities above the normal limits for our species with the use of drug therapies and medical procedures. It is very likely that we will soon be able to enhance many more of our capacities above those limits. The prospect of enhancing human mental and physical capacities above normal upper limits has been embraced by several leading bioethicists and philosophers. However, a number of very prominent intellectuals, including Michael Sandel, Leon Kass, Jurgen Habermas and Francis Fukuyama have raised a series of concerns about human enhancement. Some of these are prudential, but others are presented as moral concerns. It has been claimed that it is morally wrong to alter human nature, that to attempt to do so amounts to "playing God" and is a symptom of our hubris; and that we should not seek mastery of ourselves and others. It is also argued that we should heed our emotional responses to new biotechnologies, as some emotions, such as repugnance can be a source of moral wisdom. (p. v)

It is beyond the scope of the present book to delve further into debates about definitions of human enhancement and their bioethical and philosophical implications. For more discussion of definitions, see Gyngell and Selgelid (2016) who have addressed these matters. They argue that there are at least six ways of defining enhancement and note the implications of each.

Garland-Thomson (2015), from a disability studies perspective, suggested that a commitment to maintaining human diversity might allow greater consensus on what the possibilities and limits of human enhancement should be. She rightly noted that an important consideration is regarding how differences such as disability can be valued for what they add to people's lives and to society. In her words:

> I contend that we should conserve the human variations we think of as disabilities because they are essential, inevitable aspects of human being and because these lived experiences provide individuals and human communities with multiple opportunities for expression, creativity,

resourcefulness, relationships, and flourishing. The persistent forms of human biodiversity we consider disabilities witness sturdiness more than fragility, interdependence more than isolation. To live a nondisabled life does not secure these positive qualities, nor does living life as disabled preclude such beneficent prospects. In other words, it is not disability status or experience that determines quality of life, opportunity, happiness, or contribution to community. We are all made from flesh, blood, bone; this enfleshment sets the limits and possibilities of our existence. This shared humanity invites us to recognize what we gain from disability and what we lose when we exclude it from our shared world. (p. 13)

I concur with Garland-Thomson that maintaining human diversity should be central in our thinking about human enhancement. As we have learned in the case of global warming, destruction of other species, and fragile ecosystems, all of which diminish the planet's capabilities to support human and other life, threatens to kill life as we know it on this planet. A diversity approach to human enhancement can help to remind us just how much is at stake in what we do now and in the future.

Dilemmas of Human Enhancement

Before discussing the commodification of human embodiment, I want to consider some of the other dilemmas associated with human enhancement. One, as noted earlier in the chapter, is that there are multiple definitions of what constitutes human enhancement. Giubilini and Sanyal (2016) offered a relatively comprehensive definition: "In bioethics the term 'human enhancement' refers to any kind of genetic, biomedical or pharmaceutical interventions aimed at improving human dispositions, capacities and well-being, even when there is no pathology to be treated" (p. 1). Efforts to enhance human abilities and even emotions are controversial. On the one hand, there is public support for enhancements that help to make an injured and impaired body "better" through enhancements such as bionic limbs. The following excerpt from a recent *Science Daily* article ("Human Enhancement," 2019) characterized the practice and debates about it in this way:

Today's new human enhancement technologies are mainly used restoratively following an accident, illness or handicap of birth. A recent US study led by Debra Whitman (published in Scientific American) has shown that these restorative technologies receive near-universal approval from the general public: 95% of respondents support physical restorative applications and 88% cognitive restorative applications. This percentage drops to 35%, however, when the subject turns to interventions intended to upgrade a physical or cognitive ability with the sole aim of boosting performance. Why? "Because you're touching on the very essence of humankind, and that raises an avalanche of ethical questions," says Daphné Bavelier, professor in the Psychology Section in UNIGE's Faculty of Psychology and Educational Sciences. (para. 2)

The same article pointed out that human enhancement also has the potential to curb both human autonomy and diversity. Moreover, it is likely to create situations in which people able to afford and receive enhancements have advantages (e.g., in terms of access to particular jobs) at the expense of those who have been unwilling or unable to have their abilities enhanced. The article outlined an example of this:

Autonomy is making one's own informed decision about how to lead one's life, without being coerced by another person. It follows that an individual may choose whether or not to upgrade his or her faculties. "But," suggests professor Bavelier, "that can quickly lead to certain aberrations. If a military pilot has their eyesight enhanced, it's possible that this improved visual acuity may become obligatory to do the job. So, someone who wants to become a pilot but doesn't want to be operated on would automatically be eliminated from the profession." (para. 4)

It is important to acknowledge that individual decisions about whether to enhance a person's abilities have implications for both society and humanity. Consider the following scenario:

"If parents were able to choose certain traits for their baby, such as muscle strength, eye color or intelligence, this could have a severe impact on human diversity," says Simone Schürle, a professor in the Department of Health Sciences and Technology at ETH Zurich. "Certain trends might favour particular traits, while others might disappear, and that would tend to reduce genetic variability." And yet, each set of parents would only be choosing traits of a single baby. (para. 5)

The modification of life itself through human enhancement can thus be read as a scientific and technological advance but also a development that has potentially horrific consequences such as diminishing the variability of the human gene pool. It promises hope to those with diminished abilities but also has the potential to encourage the further development of a global order in which pressures to become more and more "able" and enhanced are intensified. This, in turn, raises issues such as global class inequalities in who can and cannot be "enhanced." In a world in which cancer treatment is often unavailable in low- and middle-income countries and in which neo-liberal policy regimes are harshly cutting back on disability benefits and assistance in accessing aids such as mobility or other devices, it is the rich that have distinct advantages in securing their own enhancement of abilities.

Human enhancement can thus be seen as part of the "slippery slope" of ableness as a regime of power and oppression. They are a set of practices that have the potential to set the bar of what constitutes able minds and bodies ever higher and, in doing so, make achieving ableness ever more difficult. Put another way, human enhancement runs the risk of runaway ableness that excludes increasing numbers of able and disabled citizens, particularly among those whose class locations mean that they lack financial and other resources that could help to secure mental and/ or physical enhancements. While a detailed review of the human enhancement literature is beyond the scope of this book, it is worth noting that disputes over such initiatives often pivot around the implications of these practices for how we deal with "human nature." Some conservative opponents argue that human enhancement goes too far in violating our "true" human nature, whereas some liberal proponents argue that enhancement has the potential to correct for advantages and disadvantages in characteristics and abilities conferred in what is sometimes called the "genetic lottery." If the latter is deemed "unfair," then human enhancement can be portrayed as "levelling the playing field." Still others object to this line of argument – noting, for instance, that it is problematic to treat the genetic lottery as a matter of social injustice when it is a natural process (see Guibilini & Sanyal, 2016, for further details).

An interesting variant inspired by commitments to human enhancement is that of the "reablement" of older people in countries such as Canada, Australia, and Denmark. This involves a treatment plan that goes beyond traditional home care services and emphasizes achieving goals such as greater independence and the ability to stay in one's home as long as possible. Like other initiatives such as workfare for disabled people, however, this is also intended to diminish the costs of caring through reductions in the hours of care needed (e.g., Aspinal et al., 2016).

Commodifying Human Embodiment

What are some of the links between human enhancement and the commodification of human embodiment? First, both share a world view in which human beings are "perfectible." This, of course, is not a new idea, as illustrated, for example, in eugenic movements that sought to intervene in human activities such as reproduction (e.g., through sterilization) of those regarded as "inferior" or defective, and in the Nazis' killing of disabled and Jewish people. Second, that there are huge profits to be made through the sale of commodities designed to enhance human embodiment. One example of this are the profits made in the pharmaceutical industry. A current report from the Centre for Americans for Progress had the following to say about drug profit. While admittedly a partisan source, there is a ring of truth to what they say (certainly compared to the former White House administration of Trump):

> President Trump has long promised to stand up to the pharmaceutical industry and lower prescription drug prices. But he has avoided taking serious action to drive down prices while at the same time filling top spots in his administration with industry insiders. This administration's culture of corruption, which continues a decades long practice of political pandering to the pharmaceutical industry, carries a real cost; Americans spent $535 billion on prescription drugs in 2018, an increase of 50 percent since 2010. These price increases far surpass inflation, with Big Pharma increasing prices on their most-prescribed medications by anywhere from 40 to 71 percent from 2011 to 2015.
>
> Moreover, pharmaceutical companies receive substantial U.S. government assistance in the form of publicly funded research and tax breaks, yet they continue to charge exorbitant prices for medications. (Meller & Ahmed, 2019, paras. 1–2)

At the global scale, profits from the pharmaceutical industry are staggering. Barton (2020) estimated that the worldwide profit from the sale of medication was US$1.3 trillion and she noted that the top ten pharmaceutical companies accounted for one-third of sales. This was worth US$392.5 billion. By 2022, the pharmaceutical industry worldwide generated US$1.48 trillion in profits (Mikulic, 2024).

Substantial profits are also being made through cosmetic surgery. In 2018, the American Society of Plastic Surgeons reported that American consumers paid more than US$16.5 billion for cosmetic surgery procedures (American Society of Plastic Surgeons, 2019).

These are some of the links between capitalism and the commodification of human embodiment. Moreover, it was argued in the previous chapter that contemporary capitalism has at its core the antagonistic relation between capital and labour. This takes the form of the violent appropriation of people's capacity to labour by capitalists. It was also argued that, especially with phenomena such as the rise of precarious employment, this antagonistic relationship places an embodied subject's continued capacity to work and to escape poverty and unemployment at risk.

There are many ways in which we are all caught up in forces promoting the commodification of human embodiment. By this I mean forces that help to transform aspects of our embodied beings into commodities for sale and consumption. A mundane example is that of the fashion industry and advertising pressures to keep up adorning oneself with the latest "trends" (at least if one is affluent enough to afford them). The fact that many of these commodities are manufactured in poorer nations of the world adds to the social injustices associated with their consumption (e.g., exploitation of poorly paid workers). We also continue to see the commodification and enhancement of people's abilities to work and consume. Consider, for example, the wide array of electronic devices now available to, in theory at least, allow humans to do more faster and more effectively (e.g., cell phones, the internet, laptops), rendering commodities such as typewriters largely obsolete. In Haraway's (1991) terms, we are increasingly becoming "cyborgs" as we adopt these facets of being part human and part machine. She described the condition of being "cyborg" in the following way:

> By the late 20th century, our time, a mythic time, we are all chimeras, theorized, and fabricated hybrids of machine and organism; in short, we are cyborgs. This cyborg is our ontology; it gives us our politics. The cyborg is a condensed image of both imagination and material reality, the two joined centres structuring any possibility of historical transformation. (p. 118)

The cyborg as an imagined hybrid state and a real material condition of being shapes our expectations of the future and the material realities of our day-to-day lives. (By this I mean that we imagine the possibilities and limits of our capabilities as future cyborg selves while also experiencing changing material conditions of life such as enhanced mobility through the use of such aids as scooters.) Moreover, with technological advances such as stem cell research and genetic testing, we are now at a point where we can modify, as Dyck (2010) usefully put it, life itself.

In becoming increasingly commodified cyborgs, we engage in more invasive modifications of the mind and body itself – modifications that are often highly profitable in our able neo-liberal capitalist world. As noted above, these include procedures such as cosmetic plastic surgery. According to the American Association of Plastic Surgeons, in 2018 there were over 15 million minimally invasive procedures performed in the US (e.g., botox injections, dermal fillers). In terms of cosmetic surgery procedures in the same year, there were 1,679,164 performed in the US (e.g., facelifts, breast augmentation) (American Society of Plastic Surgeons, 2018). The lure of plastic surgery, particularly for women, is that it promises to revive the appearance of youth and ability (e.g., to attract partners). Whether it does so or not, of course, depends on the patient and surgeon, among many other factors.

With increasingly compulsory ableness, we are also seeing a growing demand for healthy and able bodies. This finds expression in the consumption of vitamins and healthy food, fitness routines, books on health and diet, as well as gym memberships, meditation, yoga, bicycle riding, and so on. The health goals of such activities are laudable, but they are also about commodifying and profiting from desires for health and attractiveness. In the United States in 2017, there were 60.87 million gym memberships (Statista, n.d.). Between 2008 and 2014, gym memberships rose by 18.6 per cent, generating $24.2 billion in revenue – a 6.4 per cent increase over 2013 (Franchise Help, 2020). And there are, of course, downsides to this commodification. One is that most of these products, services, and activities are only accessible with relatively high incomes so that the possibility of attaining an able embodiment also becomes a class inequality issue. Another downside is that commodifying bodies in such ways renders embodied subjects who cannot satisfy these ideals as necessarily less desirable and less successful members of society. In short, by reinforcing ableness, one also tends to perpetuate intolerance for bodies and minds that differ from the norms of this and intersecting regimes of power and privilege. Hate crimes, particularly against people such as disabled and/or transgender persons, are one example of this phenomenon (E. Hall, 2019; Moreau, 2020).

In Canada, the total number of police-reported hate crimes per 100,000 population in Census Metropolitan Areas rose from 1,409 in 2017 to 1,798 in 2018 (Moreau, 2020). These figures represent only those crimes reported to police and so underestimates the extent of this problem. Further increases in hate crimes in Canada were experienced from 2020 to 2021. The number of hate crimes reported by police rose from 2,646 cases in 2020 to 3,355 in 2021. This was a 27 per cent increase (Statistics Canada, 2024). As of 2023, a task force has been struck, consisting of members of the Canadian Race Relations Foundation and the Canadian Chiefs of Police Association, to delve further into this problem.

Intolerance towards disabled and/or other vulnerable people is also reflected in more mundane ways such as discrimination and marginalization in the workplace. According to the Canadian Human Rights Commission (CHRC; 2022), 52 per cent of all discrimination complaints accepted between 2018 and 2022 were related to disability. Many of these were related to forms of disability discrimination such as diminished pay compared to able workers and failure to accommodate the disabled workers' needs to the point of undue hardship (this can sometimes be related to health and safety issues but it is usually judged by the courts in terms of undue economic hardship for the employer). Such issues have resonance in my life and experiences in academia: as discussed in chapter 3, even with overt salary discrimination ultimately avoided in my case, the career progress/merit system through which salary increments are assessed pits my performance against that of able colleagues. The result is that my performance is always judged as below par. Or as one colleague recently put it with respect to both my lengthy struggle for accommodation and current employment circumstances: "So basically you were set up to fail." This is damaging not only with respect to income and benefits such as pensions but also helps to undermine a person's sense of self-worth, job satisfaction, and belonging.

With the considerable profits to be made from providing commodities and services that aid in the pursuit of able embodiment, it is perhaps not surprising to find that some forms of such "human enhancement" take illegal, exploitative, and ethically dubious forms. I illustrate this in the next subsection with reference to the global trade in human organs for transplantation purposes. But first, I want to acknowledge how contentious the pursuit of human enhancement in general is. Freiman (2018), while recognizing concerns that mind and body enhancements may violate the boundaries of "human nature" or "being," has simultaneously argued that parents have a responsibility to enhance their children's mental and physical abilities if this allows them to "succeed" in a way comparable to their peers (e.g., through education, work, and well-being).

Gregor Wolbring, a Canadian bioethicist featured in the documentary film *Fixed: The Science Fiction of Human Enhancement* (Brashear, 2013), has taken a contrary view. He has argued that enhancements such as bionic legs for those without lower limbs (such as himself) can be expressions of intolerance for human diversity and that we ought to focus instead on celebrating the normality of doing things differently (i.e., in his case, crawling or wheeling instead of walking). This has also been taken up by other scholars such as Hansen and Philo (2007).

Hidalgo (2018) made a case for what he has called "cosmopolitan moral enhancement." By this he means:

> According to cosmopolitans, we have general duties to aid other people and to refrain from harming them simply in virtue of their humanity. Some cosmopolitans deny that ethnic, national or other social groups have fundamental moral significance and that membership in these groups justifies special obligations. Other cosmopolitans argue that even if we do have

special obligations to our co-ethnics or compatriots, our general duties to other people trump these special obligations. Cosmopolitans oppose nationalism and other forms of in-group partiality, and advocate in favor of policies that aim to protect universal human rights. (p. 172)

Hidalgo looks towards a future in which morality may be advanced through various forms of mental enhancement such as drugs or genetic modifications, encouraging greater empathy for the situations of "outgroups," particularly in poorer parts of the world, and for their needs for aid from richer countries and also curbing tendencies towards racial and other forms of intolerance towards embodied diversity (e.g., animosity towards immigrants and refugees). While it may be tempting, in these harsh right-wing neo-liberal times, to contemplate engineering emotions in this way, the dilemma it raises is whether this is going too far and threatening human diversity as well as individual freedom and choice.

Wherever we stand in such debates for and against human enhancement, this is a burgeoning field of scientific and technological advances and one that is already shaping our own and others' lives. One of the challenges we face in this context is determining whether such practices are changing human lives and societies for better or worse. To answer this question, we also need to ask who wins and who loses from enhancement. As I have argued above, unless you are part of the affluent class elite, then you are likely on the losing end of the spectrum.

In the next subsection, I turn to another set of practices aimed at "human enhancement," namely, the global trade in human organs. This trade and people's diverse places in it are also cautionary tales about the centrality of wealth and poverty in driving the acquisition, transplantation, and selling of human organs in our contemporary able neo-liberal capitalist world order. Sometimes dubbed "transplant tourism" it is one branch of so-called "medical tourism" in which more affluent recipients purchase organs (usually kidneys) from organ brokers and impoverished sellers in less affluent countries of the world.

Towards the Ends of the Earth: In Pursuit of Able Embodiment through the Illegal Organ Trade and Transplantation

I now want to turn to an arguably darker side of the quest for able embodiment in our contemporary able neo-liberal capitalist world order. This is the illegal trade in human organs (particularly kidneys) for human transplantation (for a summary of key facts about the trade, see Scheper-Hughes, 2009). These transactions involve more affluent citizens from developed nations, sometimes dubbed "transplant tourists"

(Ambagtsheer et al., 2013), purchasing organs from organ brokers – organs sold by people in poorer countries who are desperately trying to lift their families out of debt and deep poverty. Brokers profit from matching desperate patients with people who decide to sell an organ (or part of an organ in the case of livers) as the only way to meet their families' needs. Medical personnel profit from surgical and other hospital fees (e.g., blood tests matching a donor with a recipient). While organ trafficking for profit is illegal in most countries, the existing literature suggests that influential actors such as medical personnel often "look the other way" if they suspect an organ may have been obtained illegally and profit made from it (Bienstock, 2013). This contrasts with legal transplants that occur for altruistic reasons (helping a member of one's family) rather than profit.

Because organ trafficking (the selling or forced removal of an organ) is illegal and operates as part of the so-called "underground black market," it is difficult to estimate the extent of the problem. This is exacerbated by the fact that those who donate for profit frequently use forged documents to prove that they are donating to family members or other individuals for altruistic reasons (Bienstock, 2013). The existing literature has estimated that illegal kidney transplants account for 10 per cent of all kidney transplants worldwide (Hudson, 2008) and, for the organ trade as a whole, it is estimated that 5–10 per cent of these are illegal (Ambagtsheer et al., 2013). Whether or not these figures are accurate is difficult to determine at this time.

Why does the illegal organ trade thrive? Hudson (2008) answered this question in the following way:

> Globalization has brought the once rare procedure of organ transplantation, before carried out only in developed nations, to countries around the world. This set the stage for the competition between public and private healthcare entities, and in combination with surgeons willing to take risks for money and a pool of indebted citizens willing to do whatever it takes to keep their family afloat, the organ trade thrives. (para. 4)

The purchase of organs from brokers dealing with desperate patients in wealthier nations and those willing to sell an organ for profit is nothing if not controversial. Proponents of regulating the organ trade for profit argue that, with a worldwide shortage of organs, this is the most feasible way to increase supply and to ensure that organ donors are not, as currently is the case, paid very little and even nothing for their organs. They argue that this is a "win-win" situation in which the patient's life is prolonged and a family in desperate need gets financial resources (Ambagtsheer et al., 2013).

One of the problems with this argument is that what the donor is paid for his/her organ is usually not enough to lift the donor's family out of debt and poverty. Instead, most of the fees paid by patients desperate for a transplant are captured by the organ brokers and medical personnel who collude with them (Capron & Delmonico, 2015). Moreover, existing research has indicated that the organ seller often experiences poorer health after the procedure and that this limits possibilities to earn an income (Scheper-Hughes, 2005) – in short, the economic situation of the donor and their family often worsens due to organ removal, poorer health, and related incapacity to work. In some cases, sellers even die after the operation (Scheper-Hughes, 2003, p. 1646). A related problem is that there is little or no follow-up health care for those who sell their organs, which further damages their health (Hudson, 2008). In contrast, the affluent citizens who receive an organ get ongoing health care services. Moniruzzaman (2012), who wrote on organ trafficking in Bangladesh, noted that organ selling has had a wide range of negative impacts on sellers: in addition to economic and physical harm, there is psychological (e.g., depression) and social harm (e.g., stigma and avoidance). And, in a country where 78 per cent of people live on less than US$2 per day, economic desperation provides a basis for growth in organ trafficking despite such harms (pp. 70–1).

Others have argued that trafficking in human organs is morally and politically problematic and that all transplant procedures should be done for altruistic reasons among all of those involved. They sometimes criticize practices, such as in China, where organs are harvested from executed political prisoners. Although Chinese authorities dispute this, there are concerns that at least some of their "donors" are prisoners executed because they were incarcerated for conscientious objection to the Chinese political regime (Sharif et al., 2014). A related concern is that the demand for organs has been such that the demise of some of these prisoners was accelerated in order to harvest them.

Columb (2017) provided a more practical reason for rejecting a regulatory solution to organ trafficking. Examining the situation in Cairo, Egypt, where members of the Sudanese population are involved as sellers and brokers in the kidney trade, he argued that efforts to enforce a legal ban on organ trafficking have had the harmful effect of pushing the trade further underground and making it harder to monitor let alone regulate.

There are also concerns about the politically problematic privilege of being able to purchase an organ from a desperate "donor." To at least some, this is a blatant example of how a deeply unequal capitalist order exploits the poor for the lives and well-being of more affluent people

(Budiani-Saberi & Delmonico, 2008). Scheper-Hughes (2003) made the following case against regulation of the for-profit organ trade:

> Rational arguments for regulation are, however, out of step with the social and medical realities pertaining in many parts of the world where kidney selling is most common. In poorer countries the medical institutions created to monitor organ harvesting and distribution are often underfunded, dysfunctional or readily compromised by the power of organ markets, the protection supplied by criminal networks, and by the impunity of outlaw surgeons who are willing to run donor for dollars programmes, or are merely uninterested in where the transplant organ originates. (p. 1646)

Greenberg (2013) nicely summed up the debate for and against regulation:

> Western countries prohibit their citizens from selling and purchasing organs in their jurisdiction yet fail to address the issue of their citizens who purchase organs in other countries. Most scholars, such as Sheper-Hughes, who is very vocal on this issue, support this prohibition whereas others have suggested that the organ trade should be legalized and placed under state supervision to minimize exploitation and the medical dangers attendant on illegal transplants. (p. 1)

The repercussions of selling an organ are physical, economic, psychological, and social. In addition to physical problems such as hypertension and kidney insufficiency, which often precludes the heavy labour previously engaged in (e.g., construction) and loss of household income (Scheper-Hughes, 2005), sellers often experience stigma and social isolation. In Moldavia, as well as other countries such as the Philippines, some men report no longer being regarded as suitable for a marital spouse as well as being shunned or avoided (Scheper-Hughes, 2003, 1647).

There are national and international legal instruments aimed at curbing human and organ trafficking. As Capron and Delmonico (2015) have noted:

> The best-known legal instrument on human trafficking, the Palermo Protocol ... adopted by the UN in 2000 ... includes the "removal of organs" among the forms of exploitation it prohibits ... Moreover, on July 19, 2014, the Council of Ministers approved the "Council of Europe Convention against Trafficking in Human Organs" that recognizes the related crime of trafficking in organs and explicitly reaches the activities of physicians and hospitals who utilize organs obtained through payment or involuntarily. (p. 57)

Of course, international and national laws prohibiting organ trafficking are only as good as their implementation and there are many challenges in this regard. One is that it is often difficult for prosecutors to prove that an organ was obtained illegally. This is in part due to the secretive nature of the black market and the fact that enforcement bodies are sometimes given false documents and statements that portray the donation of illegally obtained organs as voluntary and altruistic (Beinstock, 2013; Capron & Delmonico, 2015). Another challenge is that the sheer economic desperation of organ sellers is such that sellers collude in providing such misinformation. And there is, of course, desperation on the part of organ recipients in wealthier nations of the Global North that, in combination with a global organ shortage, drives this trafficking on.

What Can We Learn from Organ Trafficking about Ableness, Impairment, and Disability?

There is much to be learned about ableness, impairment, and disability from the global organ trade. One lesson is that it is a way of transferring health and "ability" from the poor and vulnerable in the Global South to affluent recipients in countries of the Global North. As such, it is a global process of class exploitation and oppression. It is also often a racialized process as many organs are harvested from persons of colour for affluent white patients. It is a legacy, then, of neo-colonial relations of power. And it is phenomenon that impairs and disables organ sellers as noted above.

Organ trafficking can also be seen as a way of taking the commodification of human embodiment to its (il)logical conclusion. If we can enhance ableness chemically and on the exterior of the body (e.g., through plastic surgery), then why not also embrace modifications inside the body through transplant tourism? As noted above, some argue that it is unethical and exploitative to profit from the trade in organs, especially since it is poorer people who suffer the negative consequences of this. Others contend that regulation of the for-profit trade in organs has the potential to remedy the injustices associated with this trade, such as lack of health care for organ sellers and inadequate payments to sellers. An article in the *Washington Post* summed up the impacts on organ sellers in the following way:

> Yet the existing medical consensus prohibits the organ trade, based on the ethical view that human organs are not a commodity to be bought and sold. It is also argued that the trade is inherently exploitative, since it is the poor and vulnerable members of society who sell their organs

to the rich. Furthermore, kidney sellers receive only a small fraction of the $100,000–$200,000 typically paid by patients and rarely experience the hoped-for economic improvement. Many, in fact, suffer a deterioration of their health, which further worsens their financial problems, along with a sense of hopelessness and social isolation. (Efrat, 2016, para. 7)

The geographically uneven production of impairment is part and parcel of the pursuit of able embodiment through organ trafficking. Through negative health and economic consequences for sellers, such as diminished strength and ability to work, impairment and disability are disproportionately concentrated in countries of the Global South. Affluent organ recipients in the Global North, on the other hand, often benefit from improved health and ability post-surgically. The production of impairment and disability is, then, characterized by issues of transnational class inequality and the reproduction of neo-colonial relations of power and oppression.

The global organ trade is also a reminder that, like other forms of human trafficking, it is desperation that drives people to go to figuratively and literally "the ends of the earth" to sell and receive organs. Impoverished sellers are desperate to find ways of lifting themselves and their families out of deep poverty while recipients are desperate for another chance at life. But it is, of course, not just desperation that drives organ trafficking – it is also greed and the seemingly insatiable pursuit of wealth on the part of organ brokers, surgeons, and other health professionals.

There are also cautionary lessons from the global trafficking in human organs discussed above. One is that, in our contemporary able neo-liberal global capitalist order, we are pushing the boundaries of the commodification of the human body and mind like never before, and it is unclear where this process will end. And this is happening through class-based exploitation of the poorest and most vulnerable people in the world. In doing this we are also concurring with notions that "everything has its price," including even life itself. This is in line with the worst excesses of our able neo-liberal capitalist order, including valuing wealth over the lives and well-being of so many humans as well as other beings. We have created a world in which the truly unacceptable is constructed in practice as acceptable for the sake of profit and wealth. So, for example, and as noted earlier, we continue to fail to adequately respond to climate change as doing so would diminish profit-making in and the dominance of the fossil fuel industry. Given the facts that what is at stake is life itself and the ability of the planet to sustain lives, this is an extremely worrying situation and far too big a price to pay for continuing with "capitalism as usual" (Klein, 2015, 2019). Moreover, while wealthier countries are the greatest source of emissions, it is poorer nations of the world that are bearing the brunt of the negative impacts of climate change, thus adding to global injustice between the Global North and Global South.

Another cautionary lesson from the global trafficking in human organs is just how disposable human and other beings' lives have become. It is not only organ sellers whose lives and abilities are at risk, but, as argued in chapter 5, also those of vulnerable groups such as disabled and elderly people who, through harsh austerity policies such as cutbacks in support and services, are having their abilities to survive and their lives put increasingly at risk. I have argued that this amounts to what we might term a "violent cleansing," in which slow and fast deaths occur because of cutbacks and yet are not named for what they are.

Conclusions: From the Commodification of Human Embodiment to Legal Struggles for Disabled People's Human Rights

I have argued, in this chapter, that the commodification of human embodiment itself and enhancement of mental and physical abilities is proceeding apace in our able neo-liberal capitalist world. And with medical advances, such as the spread of transplantation procedures and genetic testing, we are increasingly able to manipulate life itself. This is both an awesome and troubling power. Awesome in that we in principle have unprecedented ability to modify our embodiment and lives, and troubling in that it is unclear whether we will use this ability to create a humane world or one that is oppressive and exploitative as well as resolutely capitalist in nature.

In the next chapter, I go on to consider disability human rights and struggles to realize those rights in practice. I argue that international and national human rights efforts have symbolic and political importance but also have significant limitations as the legal means to prevent the marginalization and exclusion of disabled people. This is evident in relation to my own experiences of struggling for accommodation at my academic workplace, as recounted in chapter 3, and in relation to efforts to realize disabled people's human rights in the Global South. Through an examination of disability human rights law in Canada, the United States, and Guyana, and at the international scale, I illustrate why we must go beyond the limits of law if we are to empower disabled people and combat the (re)production of impairment and ableness.

7 Beyond the Limits of Law: Disability Human Rights Law and the Struggle for Sociospatial Justice

I have argued in chapters 5 and 6 that we need to consider how our able neo-liberal capitalist order perpetuates impairment and disabling conditions of life. I now turn to another national and transnational realm through which disabled people and other "anomalous" embodied subjects have struggled to advance their inclusion and well-being: that is, national and transnational justice systems. This is an important part of the global context in which disabled and (temporarily) able people are living out their lives and hoping for a better future.

Law is a complex terrain of struggle in our capitalist societies. It holds out the promise of greater social justice and advances in human rights that have the potential to empower relatively marginalized groups such as disabled citizens. And yet the implementation of law is often fraught with difficulties such as inadequate funding, backlogs of court cases, and government inaction on such issues. This chapter begins to explore the contributions and limits of disability human rights law as a means of empowering disabled people.

Struggles to advance human rights have been central to efforts to challenge the oppression and marginalization of vulnerable groups in nations around the world. Examples include struggles to secure the vote for women in the late nineteenth and early twentieth centuries, to end slavery in countries such as the United States and Canada, to ongoing efforts to combat racial discrimination and discrimination against people who identify as non-heterosexual or non-able, in the twentieth and twenty-first centuries. Although earlier struggles were not necessarily framed in terms of universal human rights, as they are today at the global scale, the implicit message was clearly that embodied traits or characteristics such as gender or race should not be bases of exclusion.

This chapter explores some of the ways in which disability human rights law has helped to advance, at least in principle, the legal entitlements of diverse disabled people. But it also offers a critique of the "limits of law" in securing their rights and well-being. Drawing on the conceptual framework developed in chapter 2, the account of able neo-liberal capitalism provided in chapter 5, and using examples from the United States, Canada, and Guyana, as well as the international human rights system, the chapter argues that worsening material conditions of life for the majority of people with impairments and illnesses makes realizing human rights in principle and in practice at odds and that this can be seen, at least in part, in the geographically uneven outcomes of efforts to implement disability human rights law. This chapter is not intended to directly compare disability human rights law in these contexts. Rather, the intent is to flag some of the legal and other conditions of life that make it difficult to claim disability rights in practice. As such, this discussion should be viewed as partial and preliminary and as needing to be supplemented by more in-depth research. I conclude this chapter with reflections on moving beyond often misleading neo-liberal discourses of individual inclusion based primarily on human rights and towards a more comprehensive understanding of the material conditions of life that consign so many of the world's people to the margins of law and society. Law is not irrelevant to struggles to empower disabled members of society, but neither is it a panacea for a world in which disabled women, men, and children still all too often do not "count."

On Living at the Limits of Law: Autoethnographic Reflections

Before considering human rights struggles at the national and international scales I want to reflect very briefly on what we can learn about how law is lived from my own struggles for inclusion in the academy (recounted in chapter 3). My first reflection is that as important as human rights law and advocacy were to my struggles for inclusion in academic life, they did not and arguably cannot protect the individual taking legal action from ongoing discrimination. So, they did not ensure, for example, that the often different career trajectories of disabled scholars would be taken into account in decisions about salary and research funding. Nor did they mean that doing academic work differently, for instance through virtual conference participation, would be valued in the same way as conventional in-person presentations even though the work is at least comparable. My third reflection is that despite greater legal and policy attention to matters of diversity and inclusion, we remain a long way from shaking the ableness that is fundamental to how our societies work, including through the academy.

My final reflection is that, as Barnes and Sheppard (2019) have recently argued, we need to learn how to live the academy differently from its current neo-liberal form. This would include learning to value "slow scholarship" for the innovative research it is capable of producing, moving away from individualistic and narrow conceptions of academic "success" that value and reward those who produce the most publications and get the largest research grants and thus have the most ample paycheques. The lesson is that disability human rights law only gets us so far in challenging able neo-liberal capitalism and, if we are to go the distance in promoting more inclusive societies and spaces of everyday life, then we need to complement our human rights efforts with ongoing efforts to construct and live a different, less able, and less capitalist way of life. Before developing this argument, however, I first consider some of the contributions that human rights law has made to empowering disabled people. This is a necessarily selective review but helps to put later arguments in this chapter in context.

On the Contributions of Disability Human Rights Law I:
Examples from the United States and Canada

Historically, using human rights law to protect disabled persons from discrimination is a relatively recent development. In fact, during the eighteenth and nineteenth centuries countries such as the United States and Canada were more likely to use legal measures to exclude at least some groups of disabled people from their nation-states. In the case of the United States, immigration was relatively free and unregulated until the late nineteenth century. But following the Civil War, the US Immigration Act of 1882 "blocked (or excluded) the entry of idiots, lunatics, convicts and persons likely to become a public charge" (US Citizenship and Immigration Services, 2012, p. 3). At the global scale, Bashford (2014) noted that most if not all nineteenth-century immigration statutes included a clause barring the entry of idiots or the insane. In Canada, Rioux (2009) argued that disability was a political issue as early as the mid-1700s. This was linked to Britain's practices of shipping "undesirables" to the colonies (e.g., prisoners to Australia). She continued: "In the case of Canada, they shipped what were termed 'the poor, the indigent and the mentally and morally defective'" (p. 201). Today, the legacies of such legal restrictions live on. The Canadian Immigration Act's "excessive demand" clause with regard to health care and social services still renders many disabled people as inadmissible in practice (Chesoi & Kachulis 2020, p. 21; see also Wilton et al., 2017). The act also implicitly constructs disabled people as a "burden" without considering what they can contribute to society (Wilton et al., 2017).

As was the case in Canada, disability human rights law developed gradually in the United States and in direct response to an increasingly vocal disability movement (Neudel, 2011). From the 1970s on, a series of partial anti-discrimination laws were passed. But it was the Americans with Disabilities Act (ADA), passed in 1990, that was regarded as an "emancipation proclamation" for disabled people.

When the ADA was passed and put into effect two years later, it was heralded as a major victory for disability activists and as a statute with the potential to challenge long-standing discrimination and marginalization of the country's 43 million disabled people (Roszalski et al., 2010). Passage of the ADA was certainly a symbolic achievement in that it gestured towards the possibility of a more inclusionary and just society.

And, as the product of three decades of disability activism, it was a hard-won victory that supporters hoped would translate into empowerment for disabled people in virtually all realms of daily life (Neudel, 2011). This symbolic importance of the act was reflected in the fact that the crowd that gathered in Washington, DC, to witness its signing into law on 26 July 1990, by President George H.W. Bush, was the largest in US history (Neudel, 2011). The ADA helped to encourage other regions and countries such as the United Kingdom to pass their own anti-discrimination disability human rights laws. And, to some extent, the case law that developed in relation to the act upheld some of the key principles of the disability movement. For example, the tenets of independent living became a basis for challenging state guardianship laws that encouraged substitute decision-making by the guardian – this lost favour to supported decision-making in which a disabled individual could ask for assistance in decision-making as needed (Salzman, 2010).

But, as critics recognized, there were limits to what could be done using the ADA. Roszalski et al. (2010) pointed out that up until the act was amended in 2008, American case law progressively narrowed what counted as being disabled under the act so that it covered only a relatively small proportion of the country's disabled people. Marta Russell, an American disability activist and scholar, argued that, as a civil law and in the context of a capitalist political economy, the ADA was unlikely to remedy problems such as the exclusion of disabled people from employment and the class inequalities that resulted from that (Russell, 2002). This was because, she argued, disabled people were part of what Marx referred to as the "reserve army of labor." As such, their unemployment and underemployment helped to dampen wages and bolster profits. She also noted problems such as state governments' opposition to implementing the ADA on the grounds that changes to the status quo would be too expensive for them and for businesses. Russell (2002) concluded that without an explicit "right to a job," disabled people lacked the political and legal clout to challenge capitalists' control of access to employment. For her, then, efforts to promote equality of rights for disabled people exposed the contradictions of trying to achieve those rights in an unequal, class-divided society.

Erevelles (2016) noted that Russell's analysis of disability oppression begins with a recognition that mainstream adherence to normative ideologies and practices fuel perceptions of disabled people as being "less than human." She continued: "These ideologies have made it easy to justify inhumane policies that construct disabled people as disposable. One of these troubling ideologies is the overvaluation of work in capitalist society" (pp. 106–7). Erevelles also drew attention to Russell's critique of the welfare state as situating disability in a moral economy that refuses to guarantee an adequate standard of living. She went on to observe:

> Updating Russell's figures from a decade earlier, the monthly stipend that Supplemental Security Income provides disabled people (18–64 years) is $559.19 and for those who are over 65 years, this stipend is $433.19. In light of these sobering statistics, it seems imperative that disabled people's struggle for recognition as bona fide citizens needs to extend beyond ramps ... to dismantle ableist social ideologies and economic structures that constitute them as nonproductive citizens (a terminology very akin to the depiction of disabled people in Nazi Germany as "useless eaters"). (p. 108)

Under the act, disabled people and their legal advocates in the US court system have found it difficult to challenge discrimination in and lack of access to employment. In fact, plaintiffs in such cases were more likely to lose than win, at least prior to the 2008 ADA amendments. Colker (2010) noted that the literature on ADA cases indicated that 90 per cent of those cases were won by defendants prior to 2008 and that this often hinged on the courts' narrow interpretation of disability and who was eligible for ADA protections. For example, the Supreme Court ruling in *Sutton vs. United Airlines* determined that mitigation of impairments such as poor vision had to be considered in determining whether an individual qualified as disabled and was thus entitled to protection under the ADA. Conditions such as myopia (mitigated by glasses) or diabetes (controlled by medication) were excluded as constituting disabled status. It remains the case that vision that can be corrected by glasses or contact lenses does not qualify for disabled status under the 2008 act. The ADA amendments were aimed at restoring the broad anti-discrimination goals of the original ADA. With less emphasis on severe impairment and more emphasis on limitations in life activities, the amendments have helped to challenge narrow court interpretations of who qualifies as disabled. (See US Department of Justice, n.d., for more information on the amendments.)

Befort (2013) provided a comparative empirical analysis of judicial rulings before and after the 2008 amendments with the aim of seeing if assumptions about the likely impact of the amendments are consistent with court rulings. He reported:

> With respect to the first assumption, the data show a significant drop in employer win rates on the disability status issue from 74.4% of the pre-amendment rulings to 45.9% of the post-amendment rulings. The data also support the second assumption with an increase of summary judgement rulings on the qualified status issue [i.e., is the plaintiff qualified with or without accommodations for the job in dispute] from 28.2% in the pre-amendment claims to 47.1% in post-amendment claims. Finally, focusing on case outcomes, the number of plaintiffs surviving post summary motions increased from 32.3% in the pre-amendment decisions to 40% in the post-amendment decisions. (pp. 2069–70)

However, even if court rulings are consistent with the expected impacts of the ADAA, this does not necessarily translate into employment gains for disabled people. It is not just that legal change is often slow or that courts can become inundated with cases, it is also that, in practice, employers continue to discriminate against disabled workers in their hiring and retention and that much of this goes on beyond the reach of the justice system. It is thus not surprising that Bonaccio et al. (2020) estimated that for the United States only one in three individuals with disabilities were employed or 34.9 per cent of disabled people. This compared to the 79 per cent employment rate for persons without disabilities.

The examples discussed above help to illustrate how fragile disabled people's rights remain in the United States, despite the passage of the ADA. Concerns have also been raised regarding the prolonged failure of the US Senate to ratify the UN Convention on the Rights of Persons with Disabilities (CRPD). Some argue that this has detracted from America's international leadership on human rights issues and limited opportunities to refine the ADA and ADAA in line with the more comprehensive strategies for rights and inclusion envisaged in the CRPD (Kanter, 2019).

In Canada, legal protections against discrimination on the prohibited ground of disability developed gradually through the adoption of provincial and federal human rights codes and the associated adjudication of complaints, case law as well as constitutional change. Ontario was the first province in the country to pass a human rights code on 15 June 1962. This code prohibited discrimination "in signs, services, facilities, public accommodation, employee and trade union membership on the grounds of race, creed, colour, nationality, ancestry and place of origin" (Ontario Human Rights Commission, 2020, p. 12). From 1975 to 1981, the Ontario Human Rights Commission (OHRC) launched a review of the code. This was a time in which citizens were becoming more vocal in their rights claims in the wake of the civil rights, women's, student and gay liberation movements, and growing demands to recognize the plight of minority groups such as disabled people (Tunnicliffe, 2006). As Tunnicliffe (2006) explained, this was a time of increases in immigration to the provinces, giving rise to changes in the ethnic composition of the population, particularly in Toronto. It was also a time of greater representation of women in the workplace:

> From 1971 to 1975, Ontario saw a sharp rise in immigration from South Asia, the West Indies and the Caribbean, China and Hong Kong. The largest impact was felt in urban centres, particularly Toronto, where, by 1976, 60% of residents were of non-Anglo Saxon origin. This changing ethnic composition also brought a greater representation of various religions and creeds into the province. At the same time, a dramatic increase in the number of working women altered the dynamics of the workplace. Together, these changes gave rise to a larger and more visible minority in Ontario that interacted on a daily basis with the traditional white male majority in areas such as employment, accommodation and services. More interaction led to greater possibilities for incidents of discrimination, and victims often sought protection in the Human Rights Code. (p. 499)

Tunnicliffe also noted that although informal inquiries to the OHRC about possible complaints almost doubled and the number of cases of complaint also increased, there was minimal increase in funding which made it very difficult, if not impossible, for the OHRC to deal with inquiries and complaints in a timely way. Moreover, women and gay rights activists as well as disabled people criticized the code for not providing sufficient human rights protection for them. Women pointed to, for example, the failure to recognize sexual harassment as a form of discrimination they often faced. For gay rights and disability activists, the code failed to recognize rights to freedom from discrimination on the basis of sexual orientation and disability.

As Tunnicliffe (2006) reported, the review of the code by the private citizens appointed as commissioners to lead the proceedings, including public hearings, culminated in the preparation of a report with recommendations for change. The report, entitled *Life Together*, recognized inadequacies in the existing code such as the fact that the addition of "age" to the code, in 1972, only recognized those who were aged 40 to 65. The report suggested widening the definition to those aged 18 years or older. The report also recommended broadening the list of prohibited grounds of discrimination: adding marital status, family relationship, physical disability, criminal record, and sexual orientation (Tunnicliffe, 2006).

In an effort to broaden legal protections against discrimination, Canada passed a federal Human Rights Act in 1977. This act defined prohibited grounds of discrimination on bases of "race, national or ethnic origin, colour, religion, age, sex, sexual orientation, gender identity or expression, marital status, family status, genetic characteristics, disability and conviction for an offence for which a pardon has been granted or in respect of which a record suspension has been ordered." This was followed, in 1982, by the addition of the Charter of Rights and Freedoms to the Canadian Constitution. Section 15 of the Charter came into effect in 1985 and prohibits discrimination on grounds such as gender, ethnic background, and disability. Charter rights are limited, however, to cases falling under federal jurisdiction. In addition, provincial human rights codes established provincial human rights commissions to adjudicate complaints about discrimination on grounds such as disability. Provinces such as Ontario have introduced accessibility laws (Accessibility for Ontarians with Disabilities Act). And most recently, in 2019, the federal government has passed an Accessible Canada Act that, along with the Employment Equity Act, expands the Commission's mandate with respect to human rights violations (Canadian Human Rights Commission [CHRC], 2019).

In 2019, 58 per cent of the complaints dealt with by the CHRC pertained to discrimination in employment. This was followed by 36 per cent of complaints regarding services. Forty-six per cent of complaints were about the private sector and 43 per cent were about the federal government. Fifty-two per cent of complaints were disability-related (CHRC, 2019). Interestingly, over one half of disability complaints (53 per cent) were related to mental health, suggesting that this is a particularly challenging area with regard to protecting the substantive legal rights of people with disabilities (CHRC, 2019).

While the Canadian human rights system may lack some of the symbolic significance of a more all-encompassing statute such as the ADA, there have still been notable advances in disabled Canadians' human rights. Supreme Court rulings have upheld the Charter rights of deaf people to sign language interpreters in health care settings. In the Latimer case, in which a father killed his disabled daughter in response to what he perceived as her suffering, the Supreme Court found him guilty of murder, upholding his daughter's right to life and security of the person (see Stienstra, 2011, for details). The duty to accommodate disabled workers' needs has been upheld in the courts, the only caveat being that the accommodations do not cause undue economic hardship for the employer or pose a health and safety concern (the former is something that is difficult to prove, particularly in the case of larger organizations) (Barnett et al., 2012; Lynk, 2008).

But, as is the case in the United States, there are limits to what the law can do. Elsewhere, I pointed to problems such as cutbacks to federal and provincial human rights commissions that have hampered their capacity to adjudicate complaints (see Chouinard, 2010). Also, as a system driven by individual complaints, legal change to uphold disabled peoples' rights is slow and often expensive. Some legal programs, such as the Court Challenges program, which helped to support legal action that was deemed likely to transform existing law, have been cut (Stienstra, 2011). The Court Challenges program, discontinued in 2006, was not reinstated until 2017. And it took protest, legal action, and an out-of-court settlement with the federal government to bring this about (Government of Canada, Canadian Heritage, 2023). The federal government claimed that the revised program reflected its commitment to diversity and inclusion.

In Ontario, legal clinics that help to support poor and disabled citizens in claiming their legal rights, have also seen funding cuts. Currently, funding to legal aid clinics has been cut by 30 per cent or $133 million, although the provincial government has moved away from planned future funding cuts (Jones, 2019). It is also unclear that the adjudication of human rights complaints has made way for concrete improvement of disabled people's lives. For example, Canada fares better than the United States in terms of the rate at which disabled citizens are employed (49 per cent compared to 34.9 per cent), but there is still a 30 per cent difference between the employment rate for disabled people and the 79 per cent employment rate for non-disabled people (Bonaccio et al., 2020).

Challenges in realizing disabled people's human rights in practice in affluent nations such as the United States and Canada remind us not only of the fragility of the human rights system but also of how much discrimination continues to occur beyond the reach of the justice system. The large proportion of CHRC cases attributed to workplace discrimination against persons with mental health conditions in Canada likely represents the "tip of the iceberg" in comparison to the many incidents that never are adjudicated through the justice system.

While we still need to learn more about the barriers to justice faced by disabled people in different nations and regions of the world, evidence suggests that these are formidable even in affluent nations such as Canada. A case in point is Gibson and Mykitiuk's (2012) study of challenges experienced by disabled women trying to access health care and supports in Canada. Drawing on focus groups conducted with a total of seventy-four women with diverse impairments, they found that one of the barriers to accessing health care and support was the complicated "patchwork" of services disabled women had to navigate. Lack of communication between service agencies added to the women's difficulties in navigating the system. Disabled women also reported numerous instances of discriminatory practices that diminished their access to health care and support:

> Participants discussed multiple forms that limited their access to health services. These included physically inaccessible services, inferior care for those receiving social assistance and apparent gender-based differences in health care encounters. The CRPD [UN Convention on the Rights of Persons with Disabilities] guarantees persons with disabilities the same range and quality of health care as provided to other persons, yet the women discussed numerous instances of difficulties accessing preventative and other health care services. (Gibson & Mykitiuk, 2012, p. 115)

Thus far we have considered some of the challenges in realizing disabled people's human rights in practice in countries of the Global North (i.e., the US and Canada). In the next section I turn to the international disability human rights context and its overarching treaty, namely, the UN CRPD. While the CRPD can be celebrated as putting comprehensive disability rights on the international legal and political agenda, its interpretation and implementation are paradoxical and contested. This has led to difficulties in determining how to ensure that the rights of all disabled people are protected under the CRPD's auspices.

On the Contributions of Disability Human Rights Law II: The UN Convention on the Rights of Persons with Disabilities

When the United Nations General Assembly (GA) adopted the UN Convention on the Rights of Persons with Disabilities on 13 December 2006 (Kayess & French, 2008), it was the first international human rights treaty of the twenty-first century. It was also met with great enthusiasm, with eighty-one nations plus the European Union signing it at its opening ceremony in 2007. This was a record for any international human rights treaty. An optional protocol established a Committee on Disabled Persons Rights to monitor progress on protecting disabled people's rights. The Committee has enjoyed widespread support with 172 countries acceding to it as of 2019. The CRPD was meant to complement existing global human rights instruments, including the Convention on the Rights of the Child (1989) – the only convention to explicitly claim rights for disabled children, albeit this was mentioned only once. As Kayess and French (2008) explained:

> The GA mandate under which the CRPD was developed stipulated that the negotiating Committee was not to develop any new human rights but was to apply existing human rights to the particular circumstances of persons with disability. Accordingly, the Chairman of the negotiating committee has conceptualized the CRPD as "an implementation convention"; one that sets out a detailed code [for how existing rights] should be put into practice with respect to persons with disability ... Given that the raison d'être for development of the convention was that existing human rights instruments have failed persons with disability, to say the very least, it is paradoxical to propose that these instruments nevertheless provide the necessary scope or content to derive a blueprint that will secure their [disabled people's] rights in [the] future. However, despite the logical incoherence of this proposition, this was the unchallenged political/ administrative framework within which the CRPD was developed. (p. 20)

The CRPD was developed over a five-year period through an ad hoc committee with representation from UN officials, nation-states in the Global North and Global South, and non-governmental organizations (most but not all disability organizations). Adopted in 2006, the CRPD came into force in 2008 (Harpur, 2012). The Mexican government spearheaded the campaign that led to development of the CRPD and its argument that action on disability was a development issue missing from the UN Millennium Development Goals was important in winning support for the initiative by nations in the Global South. While scholars and activists often disagree on the significance of the CRPD as a basis for protecting disabled people's human rights (see Chouinard, 2018), there is no doubt that its aims are ambitious. As Lord et al. (2010) observed:

> Articles 10 through 30 enumerate the specific substantive obligations of the Convention and in keeping with the intention of the drafters that the CRPD provide comprehensive coverage of human rights as applied within the context of disability, the provisions cover civil, political, economic, social, and cultural rights. The rights are not structurally separated

into different parts of the treaty, nor does any provision indicate what articles fall into what category of rights. Rather, Article 4 clarifies that economic, social, and cultural rights are subject to progressive realization. Substantive provisions provide coverage of the right to life, freedom from torture and other forms of abuse, the right to education, employment, political participation, legal capacity, access to justice, freedom of expression and opinion, privacy, participation in cultural life, sports and recreation, respect for home and family, personal integrity, liberty of movement and nationality, liberty and security of the person, and an adequate standard of living. The provisions concerning living independently and in the community, personal mobility, and habilitation and rehabilitation represent novel formulations that nonetheless do connect to other more familiar human rights. (p. 569)

The Inter-American Commission on Human Rights and the Inter-American Court of Human Rights are the two key entities of the Inter-American System for the Protection of Human Rights (the System), both of which were established well before the 2006 CRPD. The Inter-American Commission on Human Rights (the Commission) was established in 1959, and the Inter-American Court of Human Rights (the Court) was established in 1978 and began deliberations in 1979 (Chouinard, 2018). This was followed by the adoption, in 1999, of the Inter-American Convention on the Elimination of All Forms of Discrimination against Persons with Disabilities. This convention requires member states of the Organization of American States (OAS) to take action to protect the rights of disabled people in areas such as employment, to work to promote independence and quality of life, and to raise awareness of disability (OAS, 1999). The Court appears to have been especially effective in challenging the murders/disappearances of opponents of authoritarian regimes in countries such as Guatemala. This is also the System through which instruments such as the CRPD are assessed and applied.

As I have noted elsewhere (Chouinard, 2018), the work of the Commission and of the Court has been hampered by long-term underfunding, a backlog of cases, and difficulties such as not being able to fund Commission fact-finding missions to countries under its jurisdiction. The System, funded through the OAS, receives far less funding than its European counterpart, and the US government has recently moved to cut funding to the OAS because of perceptions that member states are promoting abortion rights. This may further diminish the financial resources available to the Commission and the Court and may detract from the realization of disabled people's human rights in practice in the

Americas. Another challenge is the low rate of compliance with the System's decisions (Dulitzky, 2011). As Dulitzky (2011) explained, limited funds have forced the System to seek additional funding elsewhere:

> One would hope that the OAS would be able to finance its activities adequately, particularly in the realm of human rights protection. In reality, however the total budget of the Commission and the Court that is meant to finance all of these organs' activities, including the processing of the amparo [protection], is less than 10% of the total budget of the OAS. This has forced the Commission and the Court to rely on voluntary financial contributions from some member States and from various countries outside the region to finance their activities. For example, the Commission depends on the European Union to be able to attend [to] its backlog of thousands of petitions that have been delayed due to the fact that the OAS cannot provide sufficient personnel to process them. The Court depends on external funding to be able to hold hearings outside of its headquarters as a way of bringing the [S]ystem closer to societies in the region. (pp. 8–9)

The Commission's mandate is broad and includes reports on human rights in OAS member states, as well as receiving and assessing the merits of petitions claiming human rights abuses. It also includes efforts to negotiate a "friendly settlement" between the state in question and the petitioner. It has the power to make recommendations to member states on matters of human rights abuse although, unlike the Court, these recommendations are not binding. The Court, as the legal arm of the system, has the power to adjudicate cases referred to it by the Commission. However, some countries have not recognized the jurisdiction of the Court over human rights matters. This is the case, for example, with Guyana and some other Caribbean nations (e.g., Trinidad and Tobago) (Shaver, 2010).

As scholars such as Meekosha and Soldatic (2011) have argued, framing disability issues as matters of human rights has been symbolically and politically important, for instance in fostering new global alliances such as the International Network of Women with Disabilities. However, at the same time, applying a human rights framework developed in the Global North to the situation of disabled people in the Global South is not without its problems and limitations. It is a contemporary legacy of a violent colonial past that continues to hamper social development in nations of the Global South. It also, as Meekosha and Soldatic have noted, does not address issues such as the need for a redistribution of wealth between the North and South and within nations of the South that are preconditions for the realization of disabled people's rights in practice. In their words:

> We argue that in the global South much of the impairment or harm is a result of the legacy of invasion, colonisation and globalisation. This legacy has left many disabled people in the global South living in dire conditions of poverty. Given the close connection between poverty and disability, it could be argued that a redistribution of power and wealth both between rich and poor countries and within poor countries could have more impact on the lived experience of disabled people in the global South than would human rights legislation. Human rights instruments do not address issues of the distribution of wealth and power, and wealth has not historically been redistributed without struggle on the part of the powerless. Can supranational human rights instruments lay the groundwork for these bigger struggles? (p. 1389)

To begin to answer this challenging and pressing question, I now go on to consider some of the tensions and contradictions of the CRPD as a basis for protecting disabled people's rights transnationally. Then, in the following section, I consider some of the challenges of promoting disabled persons' rights in the context of the Global South. Here, I draw on insights from research conducted in Guyana.

One of the tensions in implementing the CRPD has been interpretations of the treaty that, paradoxically, put some rights at risk. This has been the case, for example, with respect to mental and legal capacity in decision-making. The CRPD envisages societies in which disabled people are assumed to have the same mental and legal capacity to make their own decisions with respect to matters such as health care treatments as able-bodied citizens. A key implication of this is that substitute decision-making should be abolished (e.g., such as guardianship laws and involuntary hospitalization).

In 2014, the Committee on the Rights of Persons with Disabilities, the UN body that oversees implementation of the CRPD, issued a general comment on Article 12 and other articles related to mental and legal capacity. The committee stressed the importance of upholding equality in assumptions about mental and legal capacity. What it failed to address were situations in which medical and other evidence proved that a person did not have the capacity to make some decisions. Some argue that this needs to be reassessed (e.g., Freeman et al., 2015). Lack of clarity on such matters has led countries, including Canada, to ratify the CRPD with reservations. Canada's 2010 statement reads as follows: "To the extent Article 12 may be interpreted as requiring the elimination of all substitute decision-making arrangements, Canada reserves the right to continue their use in appropriate circumstances and subject to appropriate and effective safeguards" (Dufour et al. 2018, p. 810). There are tensions, then, between how disabled people's rights are envisaged in the CRPD and the implementation practices favoured in specific nation-states. This is a source of confusion and invalidation of some aspects of the CRPD. The latter has the potential to undermine the authority granted to the CRPD as a definitive and universal treaty on the rights of persons with disabilities.

Another barrier to realizing the promise of the CRPD to empower disabled citizens "on the ground" is the lack of data on implementation practices, programs, and outcomes in different nation-states. Finlay et al. (2020) described the situation in Canada:

> The UN Committee on the Rights of Persons with Disabilities (2017, 12) highlights concern that Canada "does not have up-to-date quantitative and qualitative data on the situation of persons with disabilities" in the concluding observations of their initial report on Canada. Without accessible, compiled program data across provinces, it is difficult to determine whether Canada is meeting its commitments to the UN CRPD. (p. 475)

Insufficient data also makes it difficult to assess the degree to which the implementation of the CRPD within nation-states is increasing the reach of law as a means of empowering disabled persons.

Concerns have sometimes been raised regarding possible tensions in the implementation of the CRPD between the universalism of conceptions of human rights and sensitivity to cultural diversity in approaches to disability. Bickenbach (2009) has taken an alternative view, arguing that the two presuppose one another. He has been clear, however, about the limits of human rights law as a means of ensuring that disabled people are empowered and included in society:

> Public recognition of the human rights found in CRPD was a majestic and magnificent accomplishment. But, when the celebration is over, it is legitimate to ask what difference the expression of human rights makes. The cynic might say that people with disabilities do not need a right to a job, they need a job, and the distance between the two may be very great indeed. (p. 1118)

Finally, the unequal distribution of wealth at the global scale impedes the realization of disabled people's rights in practice, especially in countries of the Global South. In the next section, I provide some illustrations of this with respect to disabled people's lives and rights in Guyana.

Disability Human Rights in Principle and Practice in the Global South: Lessons from Guyana

Recent years have seen critiques of accounts of disabled people's lives that have focused predominantly on the situation in countries of the Global North. In response to this, disability scholars and activists are now paying greater attention to disabled people's rights and lives in the Global South. Examples of this work include Grech and Soldatic (2016) and the establishment of the international journal of *Disability and the Global South*. Space restrictions do not allow a detailed assessment of this growing literature. Instead, this section offers some insights into why disabled people's rights are so precarious in a particular country of the Global South – Guyana.

Guyana, physically located on the north-eastern coast of South America, but considered economically and culturally part of the Caribbean, was the first nation in the region to pass a disability human rights law in 2010. This was the hard-won outcome of many years of struggle by disabled persons and disability organizations. This was followed, in September 2014, by Guyana's ratification of the CRPD (Chouinard, 2014, 2018).

Guyana's Persons with Disability Act of 2010 is a comprehensive one covering a wide range of disabled people's rights, for instance, access to education, an accessible environment including accessible transportation and health care, and freedom from disability discrimination in the labour market (United Nations Human Rights, n.d.). As such it is consistent with the aims of the CRPD. Many challenges remain, however, in realizing these rights in practice. Deep poverty is both a cause and consequence of impairment and disability discrimination. My research on disabled people's lives and activism in the country revealed that most of the more than 100 disabled people interviewed were trying to subsist on government income assistance of less than US$1 per day (Chouinard, 2014, 2018; Worth et al., 2017). A few also received small remittances from family members living abroad, but this was not a reliable source of income. Almost invariably, those interviewed noted that before the month was over their income ran out and they had difficulty accessing essentials such as medication and food and even the funds for transportation to places of health care. Circumstances such as these lead to health conditions that exacerbate impairment and illness, for instance, through malnutrition and poor mental health.

As a neo-colonial nation with a violent colonial past, violence remains endemic to Guyanese society. It takes a variety of forms, including police violence against civilians (one of the highest rates in the world), racial violence, violence towards LGBTQ+ persons, domestic violence primarily against women, reckless driving causing accidents, deaths and impairments, and more subtle forms of violence such as "hate crimes" against disabled people.

Newspaper commentators have linked high rates of exposure to violence, including frequent newspaper reports on domestic assaults, particularly against women, resulting in loss of limbs and death and contributing to high rates of mental distress and suicide. According to a 2014 WHO report: "Guyana was cited as the country with the highest suicide rate in the world – 44.2 suicides per 100,000 deaths, four times the global average" (Bishop & Rawlins, 2018, para. 11).

Experts stress that there are multiple factors behind Guyana's high suicide rate. A psychologist explained this in the following way:

> There are many reasons for Guyana's high rate of suicide, says Balogun Osunbiyi, president and co-founder of the Guyana Psychological Association, and a government psychologist. He agrees with the assessment of the National Suicide Prevention Plan issued by the Ministry of Health, identifying poverty, pervasive stigma about mental illness, access to lethal chemicals, alcohol misuse, interpersonal violence, family dysfunction and insufficient mental health resources as key factors.
>
> "All of these factors are like dominoes," says Osunbiyi. "Sometimes it's more of this, sometimes it's more of that. But they are all factors on the table." (Bishop & Rawlins, 2018, paras. 14–15)

Violence and corruption are also widespread among authorities such as the police and courts. Thompson (2019) noted that violence and corruption in the Guyana police force is so rampant that its long-term motto of "Serve and Protect" is deeply ironic. She observed:

> Earlier this year, a US Department of Justice report stated that police brutality and extrajudicial killing rates are extremely high in Guyana. This is not surprising given the colonial legacy from which the GPF [Guyana Police Force] was born. Early colonists started the policing system aimed at controlling the labour and population of emancipated slaves and indentured servants. Many were imprisoned for idleness and vagrancy. Violent force against black and brown bodies was legitimized as necessary to bring about law and order. Modern day policing has not significantly changed from this model and sees officers still performing the racist and capitalist driven work of the colonial state. (para. 2)

Thompson went on to provide examples of recent police violence, arguing that the bodies tasked with dealing with complaints, such as the

Police Complaints Board, have been "lethargic" in their responses to complaints. Some recent cases of police violence include:

> The assault the police force often carries out on the Guyanese public spans the areas of physical, sexual and emotional abuse. Their targets are often young, poor and black. Some cases that stand out in our memories are 17 year old Shaquille Grant who was shot and killed by police, teenager Colwyn Harding who alleged that he was raped with a baton by police and a 15 year old young man who had methylated spirits thrown on his genitals and lit on fire by police. Untrained on how to respond to persons who are mentally ill, officers frequently maim or kill them. (para. 3)

Police violence resulting in death or impairment is thus directed not only at able-bodied civilians but also at those disabled people made vulnerable by mental illness. Guyana's National Commission on Disability (NCD) has been holding workshops to "sensitize" employers and citizens to the rights to be protected under the Disability Act. In 2016, one such session was held with new recruits to the Guyana Police Force ("NCD Calls for Disability Act to Be Implemented," 2018) but it is difficult to say if that has helped to curb police violence against the mentally ill.

Racial tensions and violence have a long history in Guyana from the subjugation of Black slaves by white plantation owners to more recent chapters such as electoral and governance rifts between East Indian Guyanese who regard themselves as superior to Afro-Guyanese and thus better suited to govern the country (for a discussion of the latter, see Gibson, 2006).

Violence against diverse members of the LGBTQ+ community is also a serious problem in Guyana. This violence takes a wide range of forms, including harassment, discrimination in hiring and employment, rape, beatings and stonings, and even murder and police intimidation and violence (Corrico, 2012). Since the late nineteenth century, laws have remained on the books that prohibit same-sex encounters between men, gross indecency, and cross-dressing. While the latter has been struck down (as of 2018), scholars argue that this legal history continues to fuel intolerance towards LGBTQ+ persons in Guyana (DeRoy, 2018). As with police and racial violence, this form of violence is a significant cause of impairment and death.

Violence against women, particularly domestic violence, also contributes to impairment and death in the country. It is estimated that as many as two out of every three women has experienced assault (primarily by male partners) (Chouinard, 2014). Even this may be an underestimate since women in the country often do not report this violence to the police as they fear for their lives and/or don't expect justice from the police or the courts. This is not surprising given corruption in the justice system, which results, for example, in magistrates taking bribes in exchange for more lenient sentences for the perpetrators of violence against women. Having said that, the fight against corruption in the country appears to be meeting with some success. According to a Transparency International Report, which measures corruption via a Corruption Perception Index (CPI) ranging from zero (very corrupt) to 100 (very clean), Guyana's score improved from 29 in 2015 to 34 in 2016. It then held relatively steady from 38 in 2017 to 37 in 2018. In 2019, however, it began to increase again to a score of 40 ("Guyana Improves Significantly on Corruption Index," 2020).

Through life history interviews with disabled women in Guyana, we learned of the stories of several women who had become impaired because of violence or who had had violence directed against them because they were disabled. Regarding the former, five women experienced "chopping violence" at the hands of a male partner or relative. This is where women are assaulted with a cutlass (readily available for use in harvesting sugar cane). In one horrific assault, one woman came to her daughter's home to find her male partner viciously chopping at her body. When he struck the daughter's head with the cutlass, she moaned that she was dead and fell to the floor. Her mother fled downstairs but could still hear the man violently chopping her dead daughter's body. He then came after her and began to chop one of her hands. To her surprise, when she asked him to look at all the blood and to stop, he did, and she fled to a friend's house for help. Not only does she now have limited use of her hand, she has also lost a daughter in a very brutal way, which is undoubtedly deeply psychologically distressing.

Other women who shared their stories with us included two women who both lost limbs to chopping violence perpetrated by a male partner. One woman lost her right forearm, and the other woman lost both of her forearms. The latter woman, who lived in a home with no running water or electricity, was dependent on others, chiefly a niece, to get water and turn on the generator at night for light. As I've noted elsewhere, she often sat alone in her home in darkness until her niece arrived – something that was certainly frightening because of the lenient sentence (seven years) handed down to her assailant and thus the prospect of his release and possible return to further harm her. And sadly, she was revictimized in the so-called "court of public opinion" by being blamed for the violence she experienced, for instance, by being viewed as promiscuous and disobedient (see Chouinard, 2014). Another woman who was mentally ill and unable to afford costs associated with accessing health care (e.g., transportation to the public hospital in Georgetown) was beaten by her children and harassed and hit with rocks if she ventured out into her community since, as her mother put it, people knew "she was not right in the head."

Another young woman born without a foot suffered harassment and abuse from a daughter and nephew living with her in her aunt's home. They took her belongings and, if she asked for them back, she would be beaten. She was also told that she did not have rights, even to her own possessions, in the home because she was disabled. When she asked the police to intervene, they refused to help despite visible bruising on her body because of the beatings. At that point, she decided she had to leave her aunt's home for her own safety.

As also happens elsewhere, disabled women and men were vulnerable to physical and emotional abuse in institutional care settings. About mid-way through our research, we visited a health care facility for adults with severe physical impairments. The facility was in a secluded rural area north of the capital city of Georgetown. Conditions in the facility were poor with most residents lying on dirty sheets with flies buzzing overhead and only one or two ceiling fans to help cope with the heat. The facility needed a dryer to help deal with the dirty clothes and sheets. Shortly before we prepared to depart, one of the staff members approached our driver who had parked at some distance away from the facility. We learned later that she wanted to convey to him that the residents were being regularly abused. Presumably too scared to say anything before other staff she had told our driver herself. This is a sobering reminder that the abuse and debilitation of people with physical and mental impairments are often hidden from view whether this be in the context of home or an institution.

The disabled men we interviewed did not report the same degree and intensity of violence that their female counterparts did – still there was evidence of harassment and neglect. One man had had a stroke and was paralyzed on one side of his body. He spoke of being called degrading names when encountering others outside his home. Another male stroke survivor reported that his family had ceased to interact with him and assist him with tasks, such as cooking, partly as a result of fears that strokes were "catching" and they too might experience one. Generally, disabled men were much less likely to report extreme violence than disabled women. One exception, however, was a visually impaired man who had lost his sight when acid was thrown in his face by members of a drug cartel.

What can we learn from such stories about law, ableness, impairment, and disability in the Global South and particularly in Guyana?

One lesson is that although human rights law is formally in place and acknowledged in principle by some, this does not necessarily mean that rights, for instance to security of the person or freedom from discrimination based on disability, are realized in practice. Both able-bodied and disabled women are targets of violence, particularly at the hands of men. Able-bodied and minded women are targets of what Puar (2017) conceptualized as the "right to maim" (see chapter 2 above). Men use forms of violence such as "chopping" to mutilate and exercise control over women's bodies and lives. And by causing impairment they help to diminish women's abilities to earn a living and survive. As noted above this can also make victims vulnerable to gossip (as well as shunning or avoidance), which blames them for the violence and injuries sustained. This, in turn, exacerbates the alienation and isolation that these women feel in their society because of male violence. Other anomalous bodies, for example those of LGBTQ+ persons, are also targets of violence, sometimes at the hands of police. One of the more heinous of these incidents was when police poured methylated spirits on a young man's genitals and set them on fire (see the quote above from Thompson, 2019). This too is exercising the "right to maim."

There are also challenges in ensuring that students with disabilities or "special needs" learn in inclusive educational settings. Educational professionals with expertise in disabled students' education needs are scarce, for instance, because of limited professional training opportunities. As of 2018, none of the eleven schools for students with disabilities in Guyana had trained "special education" teachers on staff (Cheong et al., 2018). Access to education for disabled students is also limited by attitudes on the part of parents and others that the disabled child or youth cannot learn. This results in practices such as keeping these children out of school and tucked away in the home (Cheong et al., 2018).

Guyana's National Commission on Disability, an advisory body to the Government of Guyana, called in October 2018 for the implementation of the Persons with Disabilities Act, arguing that persons with disabilities have not been benefiting from this legislation. During an NCD Media Summit, Ganesh Singh was quoted as appealing for actions to be taken so that the lives of persons with disabilities could be improved:

"All of us in the disability movement at that time [when the Act was passed] thought that we have a nice piece of legislation to protect and promote our rights but we're still almost at that stage where very little is done for us as persons with disabilities although there is a legal framework," Singh said.

The Act ... gave the relevant entities up to one decade to have implementations fully in place. However, this along with most of the other sections was never applied by the relevant Ministries.

[Singh continued:] "The Minister with responsibility for any legislation that is tabled must sign the Act into operation ... If the various Ministers chose not to develop and implement the regulations that are required; we're still left where we were." ("NCD Calls for Disability Act to Be Implemented," 2018, paras. 2–4)

Disability activists face many challenges in claiming their human rights in practice in Guyana. Lack of implementation of the act by responsible ministers is certainly one important barrier to doing so. Widespread poverty, particularly for disabled persons, is another barrier to realizing rights, for example, with respect to accessing health care and opportunities for collective organizing around disability issues. For the disabled women and men interviewed, a lack of sufficient income to afford transportation to health care centres or the purchase of expensive medicines available only from higher priced private pharmacies were ongoing problems.

Representatives of activist disability organizations in the capital city of Georgetown talked about lack of sufficient income for disabled persons to travel to organization meetings as one of the reasons why their membership tended to rapidly decline. Transportation is expensive and provided mainly by privately owned mini-buses and taxis. Drivers of the former have often refused to take disabled passengers because of concerns that mobility or other aids would take up space that would otherwise be occupied by more paying passengers and concerns about not being able to quickly replace their passengers (thus boosting profits). In Guyana, the second poorest country in South America, with an estimated annual GDP (gross domestic product) of US$4,000, the pressures to boost profits and income are intense. Exxon has discovered a vast offshore oil supply and has already begun drilling operations. Uncertainty about what this portends for Guyana's economic and social development is evident; with proponents of oil extraction seeing opportunities for a resource-driven accumulation of wealth. Others, however, point to widespread corruption as making it likely that most Guyanese people will not have a share of this wealth (Bryan, 2017; Maybin, 2019).

In 2009 protests against being denied the "right to ride" in mini-buses began. Disability activists with assistance from the NCD launched a campaign to sensitize mini-bus drivers to disabled people's rights to access transportation. Unfortunately, this effort was largely ineffective

(Worth et al., 2017), and the refusal of mini-bus drivers to stop for visibly disabled persons remains a serious problem. As a legal advisor in the Office of the Prime Minister noted, the problem was not just the need for more accessible environments but also the need to challenge negative attitudes towards those who are not "able": "It also requires a change in basic attitudes, negative attitudes that may foster stereotyping, discrimination, and prejudice towards those persons with disabilities" ("NCD Calls for Disability Act to Be Implemented," para. 10).

Rights to health care have been compromised by systemic societal problems such as poverty and violence and related processes such as the outmigration of trained health care professionals to more affluent nations such as Canada and the United States. They are motivated to leave Guyana by the promise of being able to provide more "state of the art" equipment and treatment, higher incomes, and a better quality of life. Guyana's health care provisions have been adversely affected by this loss of highly trained professionals – resulting, for example, in shortages of medical personnel in crucial specialities such as psychiatry and midwifery (Anderson & Isaac, 2007; Chouinard, 2018). Human resource policies that fail to address issues such as low worker motivation, absenteeism, and difficulties in stationing and retaining staff in remote communities also diminish the spatial reach and effectiveness of the health care system. In addition, there are issues of exclusion from any form of health care – with 12.5 per cent of the country's population having no access to any form of health care.

There are many lessons to be gleaned from Guyana's experience with disability human rights law. One is how rights are compromised by difficult social conditions such as poverty and high rates of police violence, especially against citizens with mental health challenges. Another is that no matter how comprehensive disability human rights are in legal documents, these are rights in principle and not necessarily realized in practice.

At the Limits of Law: Disability Human Rights and Struggles for Sociospatial Justice

In this chapter, I have argued that disability human rights law only gets us so far in the struggle for sociospatial justice for people with impairments and illnesses. This is especially so in the context of nations in the Global South where resources for realizing disabled people's human rights in practice are especially scarce. This is also the case with respect to groups such as Indigenous populations in countries of the Global North such as Canada.

Both the chapters in this book and work by scholars such as Erevelles (2011) point to the importance of thinking more expansively about the connections between living in a global capitalist order that prizes able bodies and minds and the forces disempowering disabled people. Such perspectives need to take the geographically uneven production of impairment itself into consideration. An important facet of this uneven geography is that the production of impairment tends to be more rampant in less affluent nations of the Global South. This reflects, in part, the hyperexploitation of workers central to global capitalism (resulting, for instance, in the collapse of an unsafe textile factory in Bangladesh in 2013). The illegal global trade in human organs is another example of this geographical unevenness – with organs harvested from poorer nations where desperation to escape poverty makes it especially difficult to resist selling one's organs. The organ trade ultimately benefits organ brokers, medical professionals, and wealthy recipients from the Global North at the expense of organ "donors." The latter, generally men and woman of colour, often experience impairment and chronic illness due to organ extraction and inadequate health care. This, in turn, has disabling consequences for the donor, such as no longer being able to perform paid work (Bienstock, 2013).

Clearly, we need a more expansive understanding of struggles for sociospatial justice, and the material conditions of life shaping them, if we are to ensure that human rights recognized formally are realized in practice. To illustrate, I conclude this section with reflections on some of the material conditions of life in Guyana that help to diminish disabled people's rights and inclusion in practice.

In the neo-colonial context of Guyana, chronic resource scarcity is a material condition of life that hampers disabled people's inclusion and well-being. This scarcity is the result of several forces, including the outmigration of trained health care workers to more affluent nations, which is fuelled not only by workers' desire for better working conditions and salaries but also by government responses to shortages of health care workers in countries such as Canada. Other highly trained professionals, including lawyers, also migrate to more affluent nations. This so-called "brain drain" has serious consequences. In Guyana, it means the loss of workers trained at relatively high cost, including specialists such as psychiatrists, who are in short supply in the country. In the case of Guyana, this has resulted in more limited training for health workers to contain the costs of replacing those who have left. As such, it can arguably be seen as an unjust transfer of badly needed human resources from nations of the Global South to more affluent nations in the Global North. Recently, a former minister of health in Guyana openly denounced this as an example of "poaching" Guyana's trained professionals (see Chouinard, 2019).

Access to mobility aids such as wheelchairs and to other aids such as prostheses is also limited. This too reflects a scarcity of trained health professionals. It also results from the high cost of importing wheelchairs or materials for prosthetic devices. The government has been unable, as a result, to provide a wheelchair or prostheses to all who need them. When wheelchairs are provided, it is usually through a charitable organization such as Food for the Poor – something that reflects Guyana's status as a relatively poor neo-colonial nation. Not only are prosthetic technicians scarce in Guyana, but the high cost of importing materials to make these devices has forced some disability organizations to investigate the possibility of using a local type of wood instead of imported materials for prostheses for arms and legs. In the case of one woman, a foreign aid agency provided her with very basic prostheses for the two forearms she lost in a cutlass attack. As the aid agency did not follow-up on her case, however, they were not repaired when they broke.

As noted above, access to transportation, in the form of mini-buses and taxis, is expensive. As it was also noted, some mini-bus drivers refuse to take visibly disabled riders for a variety of reasons. With extremely low incomes, disabled people have difficulty accessing places of health care and participating in organizing aimed at improving their lives.

As a neo-colonial nation in the able neo-liberal capitalist world order, Guyana has limited government resources to invest in its health care infrastructure and personnel given constraints such as debt repayment and a devalued currency. From 2014 to 2019, Guyana's national debt rose from US$1.58 billion to US$2.27 billion. National debt was expected to decrease to US$1.98 billion in 2020. However, projections for 2024 to 2029 have shown the national debt rising once again and the debt is expected to reach US$8.2 billion by 2029 (Statista Research Department, 2024). Access to adequate mental health care facilities remains limited and there are very few health care professionals trained in dealing with mental health challenges. This is, in part, due to high rates of emigration of trained professionals from Guyana (discussed above). Currently there are only twenty-seven trained psychologists and psychiatrists in the entire country, although this has increased from seven in 2014. This is in a country with one of the highest suicide rates in the world. There is a suicide crisis line, but this is run by the Guyana Police Force, which people likely find intimidating. In her article on mental health care, Bishop (2017) quoted a lawyer and mental health consultant, Anthony Autar, who has linked high rates of need for mental health care to the violence endemic to Guyana:

> Anthony Autar estimates at least a quarter of the population suffers from post-traumatic stress disorder. With one of the highest suicide rates in the world and stories of murder and domestic violence frequenting daily newspapers, he said, "They've been exposed to so many forms of alarming violent situations that just haven't been addressed in any point of their life." (para. 3)

With a growing incidence of domestic violence and mounting deaths and injuries from car crashes, Guyana's health care system is under additional strain. Long-standing problems in the system include shortages of trained health care professionals, unsanitary conditions in places of health care, and lengthy waits for treatment. In response to the rising number of car accidents, the government is finalizing a road safety strategy. To deal with rising domestic violence, the government has instituted a sexual violence protocol into the training of health care professionals (Pan American Health Organization and World Health Organization, 2020).

Ultimately, however, there is a need to diminish misogynistic attitudes towards women and hate towards others who embody differences, such as members of the LGBTQ+ community. Moreover, as Meekosha and Soldatic (2011) have argued, framing disability issues as universal human rights does not address social and spatial injustices arising from disparities in wealth between the Global North and Global South and within nations of the Global South. As noted above, this includes global inequalities in access to trained health workers, the associated "brain drain" of these and other professionals from nations of the Global South to those of the Global North, as we've seen in the case of Guyana, and the outmigration of women from nations of the Global South such as the Philippines to perform domestic labour in nations of the Global North such as Canada (Pratt, 1998, 2009). In addition to draining countries of domestic labourers, this movement also has emotional consequences for the women involved. Many of these women are mothers trying to provide for their families by working abroad and sending income back to family in their home countries (remittances). This takes an emotional toll on these women and their families as they try to remain connected despite long physical separations. This is what Pratt (2009) has described as the "circulation of sadness" that accompanies these transnational movements of working women.

Growing income inequality in Guyana is also a barrier to the inclusion and well-being of Guyana's disabled people. It is also a cause of impairment and disability among able people (e.g., lack of access to wholesome food and malnutrition). Constantine (2017) reported that income inequality has risen in Guyana in recent years but that the upper middle class has made some gains in income share. He argued that it is important to understand the racial composition of this group as primarily Indo-Guyanese who gained income from higher prices for agricultural goods. He also pointed out that the bottom 50 per cent lost income during the country's economic recovery or structural adjustment program through policies such as curbing government spending and privatizing government enterprises. This has resulted in job losses in the public sector dominated by Afro-Guyanese employees. In contrast, the richest 10 per cent, mostly of Portuguese origins, have maintained their income share through inheritance. He argued that, particularly in the context of Guyana, "inheritance is of some concern because it is inconsistent with widely held notions of economic justice and, in the case of Guyana, becomes alarming in the absence of any estate or inheritance tax" (p. 88). Constantine went on to note that the International Monetary Fund's structural adjustment program for Guyana caused mean income levels of the bottom 50 per cent to be cut by one-half during the 1990s, and that they have never recovered from their 1976 peak.

Perhaps a more positive lesson from Guyana's experience of barriers to the realization of human rights for disabled citizens is that disability activists have shown great determination in pressing ahead with demands for change and empowerment. Their achievements, such as passage of the Disability Act and becoming party to the CRPD, have been against difficult odds such as those discussed above.

Conclusions: Moving Beyond the Limits of Law – Struggles for Sociospatial Justice

An overarching insight that emerges from considering the limits of law as a terrain of struggle is that we need to take care not to equate disability human rights law with the achievement of sociospatial justice for disabled persons. As we have seen in this chapter, there remain many barriers to the empowerment and inclusion of disabled persons in nations of the Global North such as the United States and Canada, countries of the Global South such as Guyana, and internationally under the auspices of the CRPD. If human rights laws alone cannot deliver justice with respect, for example, to diminishing the deep poverty in which disabled people live, then what else needs to be done?

One crucial step in promoting social justice for disabled people is to encourage greater acknowledgment of the "wrongs" inflicted upon them, often by able-bodied people. Historically one of those wrongs was making public spectacles of disabled people (e.g., the use of disabled children in funding campaigns; and so-called "freak shows"; Thomson, 1999). Such practices help to portray disabled children and adults as deviant, objects of pity, and less than fully human. Another wrong has been the incarceration of disabled people in large-scale institutions such as psychiatric hospitals and prisons where they have experienced violence and abuse (Ben-Moshe et al., 2014). Efforts to deinstitutionalize disabled people from large-scale institutions have met with limited success in terms of promoting inclusion in community life. With limited access to affordable housing, a situation exacerbated by gentrification of inner-city neighbourhoods and insufficient support services, disabled people have been forced to piece together a "patchwork" of private and government venues (such as malls and drop-in centres) and services. And all too often they end up homeless and struggling to survive on the streets (Dear & Wolch, 1987; Knowles, 2005). Such wrongs violate disabled persons' rights to respect, security of the person, and access to essentials of life such as shelter.

The eugenics movement, with its emphasis on the desirability of able bodies and minds and intolerance towards those who do not embody ableness, also helped to give rise to wrongs against disabled people. These included forced sterilization of persons with intellectual impairments and, perhaps most chillingly, the murder of disabled people, who were characterized as "useless eaters," in Nazi concentration camps (Morris, 1991). These are clear violations of reproductive rights and the right to life itself.

Today, we are witnessing a resurgence of such eugenic ways of thinking. In addition to practices such as rationing health care to the more able-bodied, another example of this was a statement made by the head of the US Centers for Disease Control and Prevention in January 2022 in reference to the number of COVID-19 deaths, of which "over 75% occurred in people who had at least 4 comorbidities, so really these are people that were unwell to begin with." Some citizens interpreted this as implying that deaths from the pandemic were "fortunately" mainly affecting persons who already had co-morbidities or were unwell and/ or impaired (Diament, 2022).

Rhetorical attacks on scientists and journalists and others who deal in "facts" have also intensified in our able neo-liberal capitalist world. This takes a variety of forms, including denying the realities of climate change (Klein, 2015) and disparaging the work of climate scientists (including threats towards them on social media such as Twitter, now known as X) (Fraczakerley, 2023; Klein, 2015; Valero, 2023).

Earlier sections of this chapter have illustrated contemporary barriers to inclusion and social justice for disabled people. These also constitute "wrongs" committed against disabled persons. One of these is chronic underfunding of and cuts in funding to entities charged with promoting more equitable and just societies for disabled people. Examples include cuts in funding to legal aid clinics and human rights commissions in the Canadian context, and chronic underfunding of the Inter-American court responsible for adjudicating complaints of disability discrimination in the region under the CRPD. There are also a range of harmful practices such as discrimination based on disability in hiring and in workplaces and hate crimes against disabled people that deepen their exclusion from spaces of community life. In Guyana, these crimes were manifest in violence and verbal and emotional abuse of disabled women and men. Widespread male violence against women (e.g., chopping attacks) has also resulted in impairment and death while related practices of devaluation such as "shunning" (avoidance), name-calling, and malicious gossip that blames female victims for the violence and impairment they experience has helped to make disabled women feel as if they don't belong in their communities and nations (Chouinard, 2014).

What are some of the strategies that can be used to help make disabled people's rights less precarious and to move forward on the project of creating more inclusive and enabling societies? As earlier chapters illustrate, there is a need to tackle specific challenges such as the relatively high rates of poverty among disabled people in the Global North and Global South. A related challenge is to diminish barriers to disabled people's access to paid work and to essential services such as health care. With respect to the latter, a troubling aspect of health care that changed during the COVID-19 pandemic has been to prioritize patients who are younger, more able, and thus more likely to recover for the use of essential equipment such as ventilators (Lund & Ayers, 2020).

Ultimately, however, we need to struggle for nations and a world in which disabled people have multiple opportunities to "speak truth to power" – that is, to name the wrongs that have been done to them and by whom and to publicly seek reparations for the harms experienced. In a general sense this would be similar, in spirit, to the search for truth and reconciliation between Indigenous and non-Indigenous people in Canada. Following the many decades of struggles for the empowerment of disabled people within different nations, a global initiative that unsettles ableist privileges and practices in fundamental ways is urgently needed. It might be fruitful to explore this possibility under the auspices of the CRPD discussed in this chapter. We need to learn that the harms experienced by disabled people, such as living in poverty and being discriminated against in the workplace for being insufficiently "able," follows an oppressive logic that is not so very different from discrimination because one is a person of colour or a member of the LGBTQ+ community or embodies intersecting differences such as being a woman. Arguably, it makes as little sense to reward someone (e.g., in terms of higher wages) because they have able bodies and minds as it does to confer these and other privileges on someone because they happen to be white.

I have argued above that poverty and associated income inequality impede disabled people from accessing and being included in spaces of community life – for instance, with respect to transportation and access to places of health care. Although such rights exist in principle with the passage of the Disability Act and Guyana's ratification of the CRPD, they are not, then, being realized in practice or substantively. This is one of the contradictions and limitations of using a human rights framework to try to empower disabled people, particularly in nations of the Global South. Moreover, one of the key problems with neo-liberal discourses promising individual equality through human rights law is that they treat these rights as universal and, in doing so, fail to address deeper issues of distributional injustice in our able neo-liberal capitalist order. This, in turn, impedes efforts to understand the material conditions of life that are continuing to marginalize not only disabled people but others with "anomalous" bodies such as LGBTQ+ people. It also impedes efforts by activists and scholars to imagine futures that are less dominated by ableness, capitalism, and the uneven production of impairment and disability. Indeed, as I write these words, the global COVID-19 pandemic continues and is among those forces helping to reveal "fault lines" in our able neo-liberal capitalist world order: those forces range from the heightened material deprivation and suffering with which disabled people are being forced to contend under austerity; the vulnerability of older people to infection and death in privately run residential care homes and hospitals as able patients are prioritized; and the disproportionate number of people of colour and front-line workers contracting and dying from the pandemic virus as they go about trying to deliver essential services such as various types of care work, dispensing medications, cleaning, working in supermarkets and transporting food and other goods. These workers, often viewed as disposable and/or easily replaced, are now being constructed as essential and even heroic. While this may be a hopeful sign of the possibility for better wages and working conditions in the future, we also need to remember that many of these jobs are "precarious" (e.g., involuntary, part-time, low-wage necessitating working multiple jobs, and few if any benefits).

Hira-Freisen (2018), looking at the Canadian situation, noted that even though immigrants to Canada are entering the country and the labour force with stronger credentials (e.g., in terms of education and work experience in their country of origin), they are still over-represented among precarious workers. Women and older workers are also disproportionately likely to find themselves in precarious work situations (see McKay et al., 2012, for a discussion in the context of Europe).

If precarious work itself heightens risks of impairment and disability, for instance, as a result of stress and deteriorating mental health, it also highlights the injustices of differential access to safe and well paid work. To put it slightly differently, the injustices of material conditions of life allow some workers to work from home and heighten their chances of continuing to approximate norms of health and ableness but force others into dangerous and even life-threatening situations in their workplaces, and this can include getting to those workplaces for those who must use public transit.

The rush to reopen economies despite the pandemic, such as in the United States and Canada, also hammers home the lesson discussed in chapter 5, namely, that at the heart of the capitalist system lies the brutal exploitation of working-class bodies (and minds) for the benefit of those who own the means of production. In what I would argue is an able neo-liberal capitalist system that has run "amok," in the sense of its worst tendencies coming to the fore, we have seen the emergence of at least some authoritarian populist governments intent on placing profits, corporate interests, and wealth so far ahead of saving people's lives that the many lives lost to the pandemic can be portrayed as the largely inevitable costs of "doing business." To do things differently and better for most people requires a sea change with respect to our understanding of whose interests are worth defending. This will only happen with more effective, less corrupt, and more thoughtful and inclusive leadership than we saw, for instance, from the former US Trump administration or are now seeing from self-styled "strongmen" such as Russia's Vladimir Putin.

This is the broader context in which I consider the kinds of changes in material conditions of life needed to empower and enable people with impairments and disabilities in Guyana and to challenge the pervasive ableness at the heart of our contemporary global order.

As noted earlier in this chapter, Erevelles (2011) argued that one of the capitalist ideologies that perpetuates the oppression and marginalization of people with impairments and illnesses is the overvaluation of paid work. "Success" in this context is primarily assessed based on what work a person is able to do, how much they earn from it, and on related bases such as the promotions or awards they receive for this work. By any yardstick, this is a narrow conception of what makes a life worthwhile. I would argue that, instead, we need to nurture more expansive, inclusive, and less capitalist notions of success. This would help us value, for instance, efforts (outside of paid work) to draw on diverse people's lives to enrich social inclusion and cohesion. It would help to challenge the notion that the only meaningful contributions that people make to social well-being come about through paid work. It also has the potential to ease intolerance towards those unable to secure jobs in the capitalist labour market and combat stereotypes about older people who no longer work as being burdens to society. There is much work to be done in challenging this close linkage of ableness, paid work, and success in our societies.

In Guyana, efforts to disrupt this linkage could take the form of campaigns to celebrate and publicize contributions to sociospatial justice that occur outside domains of paid work, for instance, through the protests and struggles of disability activists. Globally, this would be a difficult task and would require multiple initiatives, including, for instance, more documentary films showcasing the various contributions that diverse people make to the well-being of others, and educational international conferences, networks, and forums oriented towards making the case that progress on inclusion and justice depends on shaking up notions that success can only be measured in the able neo-liberal capitalist terms of what people do in places of paid work.

Widespread poverty and economic inequality in Guyana and globally also need to be tackled. As I've argued in this chapter, these material conditions of life are distributional issues that bar people from claiming disability and other human rights in any meaningful substantive sense. It may be that struggles for a basic income for all, as we've started to see in nations of the Global North such as Canada, is one way forward on this. Developing more progressive tax systems that require the rich and middle class to contribute their fair share of taxes might help to give governments such as that in Guyana more resources to address the needs of vulnerable citizens. There may also be opportunities to fund programs such as basic income through deals with the corporations extracting oil from the oil field discovered off the coast of Guyana, although this has the disadvantage of perpetuating resource extraction that ideally should be giving way to green energy alternatives. This has been a controversial issue in Guyana, with President Raphael Trotman arguing that the transition to a green economy can be balanced with oil development as well as the use of carbon taxes. The leader of the opposition, Bharrat Jagdeo, has disputed this ("Any Tax like a Carbon Tax Does Not Make Sense," 2015).

Action is also needed to curb the outmigration of trained health care and other professionals if rights such as access to health care are to be realized in practice. There have been some efforts to do this through bilateral agreements that discourage nations of the Global North from actively recruiting health and other professionals from Guyana and other nations of the Global South (Chouinard, 2019).

Ultimately, challenging able neo-liberal capitalism globally will require measures to redistribute wealth from the Global North to the Global South. One way of helping to do this is if countries of the Global North agree to compensate nations such as Guyana for the violence and suffering experienced because of colonial and neo-colonial oppression, and for the unjust transfer of wealth associated with phenomenon such as high emigration rates of trained professionals. This could potentially be done under the auspices of the UN and/or a truth and reconciliation body with the authority and means to challenge more affluent countries to make amends for the pain, suffering, and harm done to poorer nations and their citizens through the exercise of colonial and neo-colonial power. Some might argue that this has already been done through mechanisms such as foreign aid but there are important drawbacks to this, such as foreign aid staff tending to dominate discussions of development agendas.

Perhaps most importantly, we need citizens of the world to revolt against a global order that values profit over people even in the context of a deadly pandemic, and that has seen its share of dictatorial would-be strongmen abusing their power and dividing their citizens instead of leading their countries towards a better, more inclusive future. We can and must do better and challenge the ableness, pain, and suffering so central to a capitalist order that works primarily for the rich. This is perhaps most clear when we are at the global peripheries of law where it is difficult if not impossible to challenge able power and privilege through human rights initiatives alone.

The next and final chapter brings this remapping of the able world in which we live to a close. In it I stress the key arguments of preceding chapters, their implications for future research, and the challenges of building a less ableist, impairing, and disabling capitalist world.

8 Conclusions: What Next? Working Towards a More Enabling and Empowering World

In this book, I have argued for a conception of ableness, impairment, and disability as matters of material social injustice in our contemporary neo-liberal capitalist order (chapters 2 and 5). Among the advantages of this perspective are that it configures the issues at stake as about more than barrier removal and/or the pursuit of individual human rights; instead, it involves the uneven production of impairment and disability, particularly in countries of the Global South, and able as well as disabled people's subjection to ableness as a regime of power and oppression. My autoethnographic reflections in chapter 3 on growing up in an ableist and misogynistic world began to illustrate, at the intra- and interpersonal scales, some of the ways in which, even as an able-bodied child, one is subject to the discipline of compulsory ableness. Chapter 4 introduced my autoethnographic reflections on living with manic depression (also known as bipolar disorder) and then considered what autobiographies tell us about how this "madness" is lived and how meaning is assigned to people and places. A key insight is that being "mad" is not a straightforwardly irrational state but involves rational as well as delusional ways of relating to places and people.

In chapter 5, I explored some key interconnections between ableness, impairment, and disability and a global capitalist order. I began by considering what Marx had to say about the violent appropriation of people's capacities to labour and how this has placed workers' lives at risk. I then considered recent developments such as the rise of precarious employment that has helped to diminish people's physical and mental well-being. I went on, in chapter 6, to examine what is sometimes referred to as human enhancement or transhumanism and struggles to attain able embodiment. I argued that a darker side of this ableness is revealed in the illegal global trade in human organs – a process that exploits and disempowers those who, in desperation, sell an organ in an, often failed, attempt to lift themselves and their families out of deep poverty.

Chapter 7 tackled the efficacy of human rights instruments in the Global North and Global South as a means of enabling and empowering disabled people. I argued that efforts to advance the inclusion of disabled people through individual, generic human rights is a legacy of neo-colonial power.

Particularly in poorer nations of the Global South, there are important obstacles to realizing disabled people's rights in practice. These include, as is the case in Guyana, underfunding of the Caribbean system meant to enforce human rights, the outmigration of trained professionals such as physicians and psychiatrists, lack of action on implementing the Disability Act by government officials, and the stigma still associated with disability and particularly mental illness in the country. These are not only matters of individual rights but also raise important distributional issues. The loss of trained professionals, for example, can be seen as transferring scarce human resources from countries of the Global South to those of the Global North. This helps to explain why, in Guyana, medical professionals, including psychiatrists, are in short supply. As you may recall from chapter 7, there are only a total of twenty-seven psychologists and psychiatrists in the entire country. This is amid a mental health crisis marked, for instance, by high rates of suicide. It is no wonder, then, that some have denounced this as a form of "poaching" that the country can ill afford.

Having recapped the arguments advanced in the preceding chapters, I now want to focus on what we can do to help build a more enabling and empowering global order and way of life.

Beyond Able Capitalism: An Agenda for Research and Action

If the violent appropriation of peoples' capacities to labour and the surplus value generated as profits lies at the heart of able capitalism, then it stands to reason that we need to work towards a different order and way of life. This is, of course, much easier to say than do! However, let me try to begin to address this.

We badly need additional research on ableness and, in particular, how it is manifest in our daily lives and why so many of us buy into it. In my academic workplace, for instance, it is taken for granted that because I am disabled and can do less than the average work performed by my colleagues that I should be penalized through ongoing poor performance evaluations and limited pay raises. There are problems, however, with this scenario – not only does it accentuate the salary discrimination I have dealt with for decades but it also fails to recognize that it is an oxymoron to compare the career trajectories of disabled and able-bodied professors. As I told one of our deans in the course of my struggle for reasonable accommodation, this was comparing apples and oranges and as such is unjust. An alternative is to follow the slogan popularized by Marx (1875/2022), "From each according to his [sic] ability, to each according to his [sic] needs!" From such a vantage point, rewarding people simply because they are able-bodied and able-minded (or white or male for that matter) makes little sense in terms of social justice. I have argued that autoethnography provides a useful method for teasing out how ableness impacts our lives. We need more autoethnographic accounts that pinpoint how all of us are impacted by ableness – but, and this is an important caveat, we first need to ensure that we have better representation of disabled people in our various workplaces so that their voices and concerns are heard.

I have also argued that the production of impairment itself needs to be part of our analysis of ableness, impairment, and disability as matters of social injustice. Especially since most people with physical and mental impairments and illnesses live in poorer nations of the Global South (80 per cent), it is important that we document material conditions of life and processes giving rise to the geographically uneven production of impairment and disability. One of those processes is the global organ trade, which, as we've seen, impairs and disables people desperate to escape poverty by selling an organ while enabling wealthy clients from countries of the Global North to get the organ transplantation they need. Another such process is the hyperexploitation of workers, for example, in the garment factories of countries such as Bangladesh and the unsafe conditions they are forced to labour in – conditions causing death and impairment, for instance, when buildings collapse or catch fire.

These conditions may be more severe in countries of the Global South but they exist in the Global North as well, as we saw, for instance, with respect to dangerous working conditions and related deaths and injuries in Amazon warehouses in the United States. We need to intensify our efforts to challenge such inhumane and exploitative conditions of working life if we hope to diminish the production of impairment and disability in the long term.

According to the International Labour Organization (n.d.-b), dangerous working conditions are on the rise. The ILO has estimated that

some 2.3 million women and men around the world succumb to work-related accidents or diseases every year; this corresponds to over 6000 deaths every single day. Worldwide, there are around 340 million occupational accidents and 160 million victims of work-related illnesses annually.

The ILO updates these estimates at intervals, and the updates indicate an increase of accidents and ill health. (para. 1)

Such statistics testify to the violent relations of exploitation of workers that lies at the heart of capitalism and the harms that this inflicts on working bodies and minds. As in our day and as in Marx's day, so long as capitalists continue to amass wealth, workers' lives are, at best, expendable. Challenging these careless practices will be neither easy nor quick. The bottom line is that more people need to be convinced that people's lives are more valuable than profits and the accumulation of wealth. It would be instructive, in this regard, to examine whether people see the COVID-19 pandemic as an opportunity to reconsider what matters most in our societies and global order. And to critically challenge governments that fail to tackle consumerism, that contribute to the destruction of life as we know it on earth, and that use harmful divisive messages to try to distract attention from the failures of right-wing populist leaders. As I've argued in various ways throughout this book, our contemporary able neo-liberal capitalist global order only "works" for a small global elite that is accumulating obscene amounts of wealth. According to a 2020 Pew Research Center report on the situation in the United States, since 1980 there has been a dramatic reduction of the share of aggregate income held by middle-class households and a sharp rise in the share held by upper-class households:

> More tepid growth in the income of middle-class households and the reduction in the share of households in the middle-income tier led to a steep fall in the share of U.S. aggregate income held by the middle class. From 1970 to 2018, the share of aggregate income going to middle-class households fell from 62% to 43%. Over the same period, the share held by upper-income households increased from 29% to 48%. The share flowing to lower-income households inched down from 10% in 1970 to 9% in 2018. (J.M. Horowitz et al., 2020, para. 14)

Zucman (2019) cautioned that estimates of the extent of global disparities in wealth are likely lower than the actual wealth gap in part because of difficulties, at least until recently, in measuring the wealth hidden in offshore tax havens:

> In many ways, however, it is not enough to study wealth concentration using self-reported survey data or even tax return data. Because the wealthy have access to many opportunities for tax avoidance and tax evasion – and because the available evidence suggests that the tax planning industry has

grown since the 1980s as it became globalized – traditional data sources may underestimate inequality. To capture the true wealth of the rich in today's world, it is key to look beyond administrative tax and survey microdata and to take instead a global perspective that attempts to capture all forms of wealth, domestic and foreign. (Conclusion, para. 1)

We are witnessing a global capital order that is imploding: with health care systems faltering (notably during the pandemic), wealth burgeoning among the capitalist elite, and austerity policies killing those most in need. If reflecting on such developments can help us to move beyond the capitalist status quo of a world that only works for the rich, then it may be the "silver lining" that we need so much in these difficult times.

After decades of work by disability scholars, we know much more about how people with impairments and illnesses are disabled than we do about ableness and the production of impairment. Still there is interesting work to be done. It would be helpful, for example, to know more about "hate crimes" towards disabled people and the extent to which these are associated with negative attitudes towards other people with anomalous bodies and minds (e.g., people of colour, LGBTQ+ people). As E. Hall (2019) noted, many such acts are relatively mundane, micro-aggressions such as name-calling and staring, but still take their toll on victims. Other research, however, suggests that the severity of violence tends to be greater in the case of hate acts directed towards disabled people (Healey, 2020; Sherry, 2016) rather than to members of other minority groups.

We also need to promote further debate about the kinds of changes/actions that are needed to help build a more enabling world for the vast majority, as opposed to an elite minority. I have suggested, for example, that the redistribution of wealth between the Global North and Global South, as well as within countries of the Global South, needs to be part of our efforts to remedy the injustices that disabled people face. As noted in chapter 7, without the resources needed to implement human rights law in principle, the disabled will not be protected in any meaningful substantive or practical sense.

Moreover, we need to act on disparities such as the flow of trained professional workers from countries of the Global South to the Global North. I have argued that this is an unjust transfer of badly needed human resources and a way in which poorer countries effectively subsidize more wealthy nations (e.g., with their shortages, for instance, of health care workers). This also works to limit access to health care in countries of origin such as Guyana. There also need to be more reparations for the harms inflicted on such nations under colonial and neo-colonial rule – such as the perpetuation of multiple forms of violence endemic to Guyana and the widespread deaths and impairments this causes.

The uneven geographic production of impairment itself also needs to be addressed as part of any meaningful strategy for change. We need to find ways to curb unsafe working conditions – conditions on the rise as noted earlier in this chapter. This will require stepped-up efforts to tackle attitudes and practices that treat workers as disposable and easily replaced. If I am right that the harms done to workers are intensifying in our able neo-liberal capitalist world order, then it is vital that we act against this. This might include further publicizing working conditions that are injurious to the well-being and abilities of workers and organizing global campaigns to boycott commodities produced by workers who are poorly paid and vulnerable to a wide range of harms in the workplace. This requires turning capitalist (il)logic on its head by putting people ahead of profits as the marker of a compassionate and more just society.

Finally, and in some ways most importantly, it is time to challenge the regime of ableness and the harms it inflicts on all of us (whether disabled or able-bodied). It is an outdated and unjust system that rewards only those who can approximate the able ideal; an ideal that, as we've seen, becomes ever more elusive as human enhancement helps to reveal. In fact, despite the potency of ableness as a regime of power and oppression, ableness is in some ways a myth, albeit a seductive one. It disciplines and teaches us to strive to be hyperable in all we do. It teaches us to value human competition over compassion and inclusion. It also teaches us that our lives must be measured by the work we do and how much we are paid for it and how much we consume. As disability scholars such as Erevelles (2011) have noted, it overvalues paid work and under values so many other aspects of our lives. It also encourages us to be complacent about the ways in which ableness becomes manifest in our lives, such as through mundane assumptions that the more work we do, the better we should be paid and, conversely, that the less we can do, the less we should be paid.

Although Fraser (2019) did not explicitly deal with ableness, impairment, and disability, her recent musings on actually existing neo-liberalisms in contemporary capitalism provided some interesting points of departure for understanding the kinds of changes needed to promote a more empowering and enabling social order. Drawing on Gramsci, she argued that we are in a hegemonic crisis and that this needs to be recognized and addressed if there is to be meaningful counterhegemonic change. Hegemony refers to "the process by which a ruling class makes its domination appear natural by installing the proposition of its own worldview as the common sense of society as a whole" (p. 9). She continued:

> Its organizational counterpart is the hegemonic bloc: a coalition of disparate social forces that the ruling class assembles and through which it asserts its leadership. If they hope to challenge these arrangements, the dominated classes must construct a new, more persuasive common sense, or counterhegemony, and a new, more powerful political alliance and a new, more powerful alliance, or counterhegemonic bloc. (p. 10)

Using the example of the United States, Fraser has argued that recent decades have seen a whittling away of the ideas and institutions through which the capitalist class continued to dominate working class people. While Trump was among those at the epicentre of the crisis in "progressive" individual neo-liberalist capitalism – a capitalism that under the weight of economic recession and now a global health crisis no longer seems to make common sense or provide a meaningful basis for a cross-class alliance – his hyper-reactionary neo-liberalism and divisive populist rhetoric complicated efforts, for instance by the Republican Party, to put a positive spin on actually existing capitalism today. In this context, disabled scholars and activists need to continue to work to forge alliances with diverse working-class people and institutions such as unions to develop a new counterhegemonic bloc and "common sense." The latter would include a shift away from individualist notions of rights and responsibilities in favour of ensuring that collective values of interdependence are seen as a better basis for collective life, and that the majority of the world's people are able to have the resources they need not just to survive but also to thrive.

Such a shift would include critical reappraisal of eugenicist rhetoric and how it informs the practices of health care professionals and insurance companies, among others. McCabe and McCabe (2011) pointed out that women in the United States who are pregnant with a fetus diagnosed with Down's syndrome experience economic coercion from insurers to terminate the pregnancy. In one case, an insurer told a woman that if she carried the fetus to term she would be solely responsible for the costs of care for the child.

There are also, at least sometimes, professional practices that pressure a patient into making a decision to terminate a Down's pregnancy – such as when a woman was told by her physician that the life expectancy for her baby would be only three years (it was actually approximately fifty years at the time when the article was written) (McCabe & McCabe, 2011). Such incidents are not as simple as passing on out-of-date information and engaging in "careless" health care practices. They speak volumes about ongoing eugenicist devaluation of the lives of disabled people. Eugenic ways of thinking are also at play in anti-immigrant sentiment in countries such as Canada, the United States, Australia, and elsewhere.

It is high time for a change from our capitalist and ableist global order and associated class injustices. It is time that our world begins to work for the majority of its people, including disabled persons. As populations age, as conditions of working life deteriorate, as austerity measures harm and even kill our most vulnerable citizens, and as we continue to witness "fault lines" in our able neo-liberal capitalist way of life (e.g., the rationing of health care in favour of more able patients), it is clear that we can no longer afford societies that value profits over people. Nor can we ignore the fact that addressing impairment, disability, and ableness is one of the most pressing challenges of our times. To build a more inclusive society will require our very best efforts to positively value non-able ways of being, to become more attuned to how ableness is manifest in our lives, and to embrace values such as interdependency and empathy over individualism, consumerism, and competition for profits and wealth.

If it does nothing else, I hope this book convinces you that we all need to be part of making this kind of sea change in our ways of life happen. In a very real sense, all our lives and futures are at stake.

Postscript: On Daring to Dream Another World

This book has addressed some of the difficult challenges we face in striving towards a more inclusive, supportive, and caring world in the twenty-first century. And it has made the case that our futures depend on doing so.

In addressing such challenges, we need to insist on our rights to dare to dream of another world. We can (and must) struggle for a world in which amassing obscene amounts of wealth is viewed as an abuse of power. We need to imagine worlds in which politicians who lie, intimidate, mock those people different from themselves, and wage war are no longer welcome as our leaders. We need to stand in solidarity with the many affected by intolerance towards embodied diversity.

In short, we need to dare to dream together of a world in which all lives (including non-human ones) matter, and embrace a truly inclusive "ethics of care."

References

Ambagtsheer, F., Zaitch, D., & Weimar, W. (2013). The battle for human organs: Organ trafficking and transplant tourism in a global context. *Global Crime, 14*(1), 1–26. https://doi.org/10.1080/17440572.2012.753323.

American Society of Plastic Surgeons. (2018). Plastic surgery statistics. https://www.plasticsurgery.org/news/plastic-surgery-statistics?sub=2018+Plastic+Surgery+Statistics.

American Society of Plastic Surgeons. (2019). Plastic surgery statistics. https://www.plasticsurgery.org/news/plastic-surgery-statistics?sub=2019+Plastic+Surgery+Statistics.

Anderson, B.A., & Isaacs, A.A. (2007). Simply not there: The impact of international migration of nurses and midwives – perspectives from Guyana. *Journal of Midwifery & Women's Health, 52*(4), 392–7. https://doi.org/10.1016/j.jmwh.2007.02.021.

Antonacci, J.P. (2020, August 14). Large scale reform needed to protect Canada's migrant workers. *Hamilton Spectator.* https://www.thespec.com/news/hamilton-region/large-scale-reform-essential-to-protecting-canada-s-migrant-workers/article_9b53e218-ae34-59c5-b9eb-6d8b89082b87.html#:~:text=In%20the%20wake%20of%20large,benefits%20and%20leave%20abusive%20farms.

Any tax like a carbon tax does not make sense – opposition leader. (2015, November 14). *Kaiteur News.* https://www.kaieteurnewsonline.com/2015/11/14/any-tax-similar-to-a-carbon-tax-does-not-make-sense-opposition-leader/.

Aspinal, F., Glasby, J., Rostgaard, T., Tuntland, Rudi H., & Westendorp, G.J. (2016). New horizons: Reablement – supporting older people towards independence. *Age and Ageing, 45*(5), 574–8. https://doi.org/10.1093/ageing/afw094.

Bains, A., & Nash, C. (2007). The Toronto Women's Bathhouse raid: Querying queer identities. *Antipode, 39*(1), 17–34. https://doi.org/10.1111/j.1467-8330.2007.00504.x.

Barnes, T.J., & Sheppard, E.S. (2019). Conclusion. In T.J. Barnes & E.S. Sheppard (Eds.), *Spatial histories of radical geography: North America and beyond* (pp. 371–416). Wiley.

Barnett, L., Nicol, J., & Walker, J. (2012). *An examination of the duty to accommodate in the Canadian human rights context.* Publication No. 2012-1-E. Parliament of Canada Research Publications. https://lop.parl.ca/sites /PublicWebsite/default/en_CA/ResearchPublications/201201E.

Barton, C. (2020, March 3). Annual revenue of top 10 big pharma companies. *thepharmaletter.* https://www.thepharmaletter.com/pharmaceutical /annual-revenue-of-top-10-big-pharma-companies.

Bashford, A. (2014). Immigration restrictions: Rethinking period and place from settler colonies to postcolonial nations. *Journal of Global History, 9*(1), 26–48. https://doi.org/10.1017/S174002281300048X.

BBC [British Broadcasting Channel]. (2021, October 8). Biden announces a $1.75 trillion spending plan. https://www.bbc.com/news/world-us -canada-59081791.

Becker, F., & Becker, S. (2008). *Young adult carers in the UK: Experiences, needs and services for carers aged 16–24.* Princess Royal Trust for Carers.

Befort, S.F. (2013). An empirical analysis of the case outcomes of the ADA Amendments Act. *Washington and Lee Law Review, 70*(4), 2027–71. https:// scholarship.law.umn.edu/faculty_articles/114.

Ben-Moshe, L., Chapman, C., & Carey, A.C. (Eds.). (2014). *Disability incarcerated: Imprisonment and disability in the United States and Canada.* Palgrave Macmillan.

Berlant, L. (2007). Slow death: (Sovereignty, obesity and lateral agency). *Critical Inquiry, 33*(4), 754–80. https://doi.org/10.1086/521568.

Bickenbach, J.E. (2009). Disability, culture and the UN Convention. *Disability and Rehabilitation, 31*(14), 1111–24. https://doi.org/10.1080/09638280902773729.

Bienstock, R.E. (2013). (Director, Producer, Writer). *Tales from the Organ Trade* [Film]. Journeyman Pictures.

Bishop, M. (2009). *Sunshine and shadow: My battle with bipolar disorder.* iUniverse.

Bishop, M. (2017, December 7). *Guyana's mental health care shortage.* Pulitzer Center. https://pulitzercenter.org/stories/guyanas-mental-health-care -shortage.

Bishop, M., & Rawlins, M.C. (2018, June 30). *Trying to stop suicide: Guyana aims to bring down its high rates.* Pulitzer Center. https://legacy.pulitzercenter .org/reporting/trying-stop-suicide-guyana-aims-bring-down-its-high-rate.

Bonaccio, S., Connelly, C.E., Galletly, I.R., Jetha, A., & Martin Ginis, K.A. (2020). The participation of people with disabilities in the workplace: Employer concerns and research evidence. *Journal of Business and Psychology, 35*, 135–58. https://doi.org/10.1007/s10869-018-9602-5.

Brashear, R.P. (Producer, Director). (2013). *Fixed: The science fiction of human enhancement* [Film]. Making Change Movies.

Bryan, A.T. (2017, November 1). Guyana, one of South America's poorest countries, struck oil. Will it go boom or bust? *The Conversation.* https:// theconversation.com/guyana-one-of-south-americas-poorest-countries -struck-oil-will-it-go-boom-or-bust-86108#:~:text=Guyana%20lays%20 the%20groundwork&text=While%20that's%20a%20fairly%20low,a%20 day%20in%20oil%20earnings.

Budiani-Saberi, D.A., & Delmonico, F.L. (2008). Organ trafficking and transplant tourism: A commentary on the global realities. *American Journal of Transplantation 8,* 925–9. https://doi.org/10.1111/j.1600-6143.2008.02200.x.

Buhariwala, P., Wilton, R., & Evans, J. (2015). Social enterprises as enabling workplaces for people with psychiatric disabilities. *Disability & Society, 30*(6), 865–79. https://www.tandfonline.com/doi/full/10.1080/09687599.2015.1057318.

Butler, R.E. (1994). Geography and vision-impaired and blind populations. *Transactions of the Institute of British Geographers, 19*(3), 366–8. https://doi.org /10.2307/622329.

Butler, R.E., & Parr, H. (Eds.). (1999). *Mind and body spaces: Geographies of illness, impairment and disability.* Routledge.

Butler, S., & Begum, T. (2023, April 24). Abuses "still rife": 10 years on from Bangladesh's Rana Plaza disaster. *The Guardian.* https://www.theguardian .com/world/2023/apr/24/10-years-on-bangladesh-rana-plaza-disaster -safety-garment-workers-rights-pay.

Cameron, K. (2020). *Being, negotiating and mending: Experiences of care in neo-liberal times* [Unpublished doctoral dissertation]. McMaster University.

Cameron, K., & Chouinard, V. (2014). On being a princess and a problem: Negotiating attitudinal barriers to chronic illness in Canadian universities. In S.D. Stone, V. Crooks, & M. Owen (Eds.), *Working bodies: Chronic illness in the Canadian workplace* (pp. 177–95). McGill-Queen's University Press.

Campbell, F.K. (2009). *Contours of ableism: Territories, objects, disability and desire,* Palgrave Macmillan.

Canadian Human Rights Commission. (2019). *Stand together: The Canadian Human Rights Commission's 2019 report to parliament.* https://www.chrc-ccdp .gc.ca/sites/default/files/publication-pdfs/chrc_ar_2019-eng_0.pdf.

Canadian Human Rights Commission. (2022). *By the numbers.* https:// chrcreport.ca/by-the-numbers.html.

Canadian Press. (2019, October 3). A list of cuts and program changes the Ford government has reversed. *CTV News.* https://toronto.ctvnews.ca/a-list-of -cuts-and-program-changes-the-ford-government-has-reversed-1.4623131.

Capron, A.M., & Delmonico, F.L. (2015). Preventing trafficking in organs for transplantation: An important facet of human trafficking. *Journal of Human Trafficking, 1*(1), 56–64. https://doi.org/10.1080/23322705.2015.1011491.

Charlton, L.I. (2000). *Nothing about us without us: Disability oppression and empowerment.* University of California Press.

Cheney, T. (2007). *Manic: A memoir*. Harper-Collins ebooks.

Cheong, K.A., Kellems, R. O., Anderson, M. M., & Steed, K. (2018). The education of individuals with disabilities in Guyana: An overview. *Intervention in School and Clinic, 54*(4), 1–5. https://doi.org/10.1177/1053451218782435.

Chesoi, M., & Kachulis, E. (2020). *Canadian citizenship: Practice and policy background paper*. Library of Parliament. https://lop.parl.ca/staticfiles/PublicWebsite/Home/ResearchPublications/BackgroundPapers/PDF/2020-64-E.pdf.

Cho, S., Crenshaw, K., & McCall, L. (2013). Toward a field of intersectionality studies: Theories, applications and praxis." *Signs: Journal of Women in Culture and Society, 38*(4), 785–810. https://doi.org/10.1086/669608.

Chouinard, V. (1995/1996). Like Alice through the looking glass: Accommodation in academia. *Resources for Feminist Research, 24*(3/4), 3–11.

Chouinard, V. (2010). Like Alice through the looking glass II: The struggle for accommodation continues. *Resources for Feminist Research, 24*(3/4), 161–77.

Chouinard, V. (2014). Precarious lives in the Global South: On being disabled in Guyana. *Antipode, 46*(2), 340–58. https://doi.org/10.1111/anti.12046.

Chouinard, V. (2018). Living on the global peripheries of law: Disability human rights in principle and in practice in the Global South. *Laws, 7*(8), 1–14. https://doi.org/10.3390/laws7010008.

Chouinard, V. (2019). Disability, migration and health in the Global South: An agenda for research and action. In K.B. Newbold & K. Wilson (Eds.), *A research agenda for migration and health* (pp. 11–28). Edward Elgar Publishing.

Chouinard, V., & Grant, A. (1995). On being not even anywhere "near the project": Ways of putting ourselves in the picture. *Antipode, 27*(2), 137–68. https://doi.org/10.1111/j.1467-8330.1995.tb00270.x.

Chouinard, V., Hall, E., & Wilton, R. (Eds.). (2010). *Towards enabling geographies: "Disabled" bodies and minds in society and space*. Ashgate.

Clarke, S., Savulescu, J., Coady, C.A.J., Goibilini, A., & Sanyal, S. (Eds). (2016). *The ethics of human enhancement: Understanding the debate*. Oxford University Press.

Colker, R. (2010). Speculation about legal outcomes under 2008 ADA amendments: Cause for concern. *Utah Law Review, 2010*(4), 1029–55.

Columb, S. (2017). Excavating the organ trade: An empirical study of organ trading networks in Cairo, Egypt. *British Journal of Criminology, 57*(6), 1301–21. https://doi.org/10.1093/bjc/azw068.

Connell, J. (2006). Medical tourism: Sea, sun, sand and surgery. *Tourism Management, 27*(6), 1093–1100. https://doi.org/10.1016/j.tourman.2005.11.005.

Constantine, C. (2017). The rise of income inequality in Guyana. *Social and Economic Studies, 66*(3-4), 65–95. https://www.jstor.org/stable/44732918.

Corbett, J. (2019, July 16). *Amid Amazon Prime Day protests, Sanders and Omar lead call for probe into "brutal and hazardous working conditions."* Common Dreams. https://www.commondreams.org/news/2019/07/16

/amid-amazon-prime-day-protests-sanders-and-omar-lead-call-probe -brutal-and-hazardous.

Corrico, C. (2012). *Collateral damage: The social impact of laws affecting LGBT persons in Guyana.* The British High Commission for Barbados and the Eastern Caribbean and the British High Commission for Guyana.

Cosgrove, L. (2000). Crying out loud: Understanding women's emotional distress as both lived experience and social construction. *Feminism & Psychology, 10*(2), 247–67. https://doi.org/10.1177/0959353500010002004.

Crenshaw, K. (1989). Demarginalizing the intersection of gender and race: A Black feminist critique of feminist theory and antiracist policy. *University of Chicago Legal Forum, 1*(8), 139–67. https://chicagounbound.uchicago.edu /cgi/viewcontent.cgi?article=1052&context=uclf.

Crooks, V.A. (2010). Women's changing experiences of the home and life inside it after becoming chronically ill. In V. Chouinard, E. Hall, & R. Wilton (Eds.), *Towards enabling geographies: "Disabled" bodies and minds in society and space* (pp. 45–61). Ashgate.

Cusi, A., MacQueen, G., & McKinnon, M. (2010). Altered self-report of empathetic understanding in patients with bipolar disorder. *Psychiatry Research, 178*(2), 354–8. https://doi.org/10.1016/j.psychres.2009.07.009.

Davidson, J. (2007). "In a world of her own...": Re-presenting alienation and emotion in the lives and writings of women with autism. *Gender, Place & Culture, 14*(6), 659–77. https://doi.org/10.1080/09663690701659135.

Davis, L.J. (2017). Introduction: Disability, normality and power. In L.J. Davis (Ed.), *The disability studies reader* (5th ed., pp. 1–16). Routledge.

Dear, M.J., & Wolch, J.R. (1987). *Landscapes of despair: From deinstitutionalization to homelessness.* Princeton University Press.

DeRoy, P., with Baynes Henry, N. (2018). Violence and LGBT rights in Guyana. In N. Nicol, A. Jjuuko, R. Lusimbo, N. J. Mulé, S. Ursel, A. Wahab, & P. Waugh (Eds.), *Envisioning global LGBT human rights: (Neo)colonialism, neoliberalism, resistance and hope* (pp. 157–75). University of London Press.

Diament, M. (2022, January 18). CDC director apologizes to disability advocates for "hurtful" comments. *Disability Scoop.* https:disabilityscoop.com.

Disability in the workplace. (2015, June 24). *Guyana Inc., 16,* 26. https://issuu .com/guyanainc/docs/guyana_inc_issue_16_final.

Diven, B. (2009). *Closing the chasm: Letters from a bipolar physician to his son.* iUniverse.

Doan, P. (2010). The tyranny of gendered spaces – reflections from beyond the gender dichotomy. *Gender, Place and Culture, 17*(5), 635–54. https://doi.org /10.1080/0966369X.2010.503121.

Dodd, S. (2016.) Orientating disability studies to disablist austerity: Applying Fraser's insights. *Disability & Society, 31*(2), 149–65. https://doi.org/10.1080 /09687599.2016.1152952.

Dolmage, J.T. (2017). *Academic ableism: Disability and higher education*. University of Michigan Press.

Dufour, M., Hastings, T., & O'Reilly, R. (2018). Canada should retain its reservation on the United Nation's Convention on the Rights of Persons with Disabilities. *The Canadian Journal of Psychiatry*, *63*(12), 809–12. https://doi.org/10.1177/0706743718784939.

Dulitzky, A. (2011). The Inter-American System of Human Rights fifty years later: Time for changes. *Revue Quebecoise de droit international*, *19*(40), 127–64. https://doi.org/10.7202/1068396ar.

Dyck, I. (1995). Hidden geographies: The changing lifeworlds of women with multiple sclerosis. *Social Science & Medicine*, *40*(3), 307–20. https://doi.org/10.1016/0277-9536(94)E0091-6.

Dyck, I. (2010). Geographies of disability: Reflections on new body knowledges. In V. Chouinard, E. Hall, & R. Wilton (Eds.), *Towards enabling geographies: "Disabled" bodies and minds in society and space* (pp. 253–64). Ashgate.

Efrat, A. (2016, December 7). Organ traffickers lock up people to harvest their kidneys. Here are the politics behind the global organ trade. *Washington Post*. https://www.washingtonpost.com/news/monkey-cage/wp/2016/12/07/organ-traffickers-lock-up-people-to-harvest-their-kidneys-here-are-the-politics-behind-the-organ-trade/.

Ellis, C., Adams, T.E, & Bochner, A.P. (2011). Autoethnography: An overview. *Forum: Qualitative Social Research*, *12*(1). https://doi.org/10.17169/fqs-12.1.1589.

Ellis-Petersen, H. (2019, December 8). "Bhopal's tragedy has not stopped": The urban disaster still claiming lives 35 years on. *The Guardian*. https://www.theguardian.com/cities/2019/dec/08/bhopals-tragedy-has-not-stopped-the-urban-disaster-still-claiming-lives-35-years-on#:~:text=Bhopal's%20tragedy%20has%20not%20stopped.%E2%80%9D&text=Activists%20allege%20that%20there%20has,to%20protect%20the%20corporations%20involved.

Erevelles, N. (2011). *Disability and difference in global contexts: Toward a transformative body politic*. Palgrave MacMillan.

Erevelles, N. (2016). Beyond ramps/against work: Marta Russell's legacy and the politics of intersectionality. In R. Malhotra (Ed.), *Disability politics in a global economy: Essays in honour of Marta Russell* (pp. 105–17). Routledge.

Estrada, M. (2009). *Bipolar girl: My psychotic self: A memoir*. Eloquent Books.

Evans, R., & Becker, S. (2009). *Children caring for parents with HIV/AIDs: Global issues and policy responses*. Policy Press.

Finlay, B., Duncan, S., & Zwicker, J.D. (2020). Navigating government disability programs across Canada. *Canadian Public Policy*, *46*(4), 474–91. https://doi.org/10.3138/cpp.2019-071.

Fitch, N. (2014, April 14). The deadly cost of fashion. *New York Times*. https://www.nytimes.com/2014/04/15/opinion/the-deadly-cost-of-fashion.html.

Flanders, S. (2020, June 26). Horrific: Texas man with disability dies after doctors refuse to treat him for COVID-19. *LiveAction News*. https://www.liveaction.org/news/texas-disability-dies-doctors-refuse-covid-19/.

Fraczakerley, A. (2023, May 14). Climate crisis deniers target scientists for vicious abuse on Musk's Twitter. *The Guardian*. https://www.theguardian.com/environment/2023/may/14/climate-crisis-deniers-target-scientists-abuse-musk-twitter#:~:text=Climate%20crisis%20deniers%20target%20scientists%20for%20vicious%20abuse%20on%20Musk's%20Twitter,-This%20article%20is&text=Some%20of%20the%20UK's%20top,by%20Elon%20Musk%20last%20year.

Franchise Help. (2020). Fitness industry analysis 2020 – cost & trends. https://www.franchisehelp.com/industry-reports/fitness-industry-analysis-2020-cost-trends/.

Fraser, N. (2013). *Fortunes of feminism: From state-managed capitalism to neoliberal crisis*. Verso Books.

Fraser, N. (2019). *The old is dying and the new cannot be born: From progressive neoliberalism to Trump and beyond*. Verso Books.

Fraser, N. (2022). *Cannibal capitalism: How our system is devouring democracy, care and the planet and what we can do about it*. Verso Books.

Freeman, M.C., Kolappa, K, de Almeida, J.M.C, Kleinman, A., Makhashivill, N., Phakahi, S., & Thorncroft, G. (2015). Reversing hard won victories in the name of human rights: A critique of the General Comment on Article 12 of the UN Convention on the Rights of Persons with Disabilities. *The Lancet Psychiatry*, 2(9), 844–50. https://doi.org/10.1016/S2215-0366(15)00218-7.

Freiman, C. (2018). Why parents should enhance their children. In J. Flanigan & T.L. Price (Eds.), *The ethics of ability and enhancement* (pp. 155–71). Palgrave MacMillan.

Frketich, J. (2020, February 10). "It scares me to death," says Hamilton PSW about severe shortages of long-term care staff. *Hamilton Spectator*. https://www.thespec.com/news/hamilton-region/it-scares-me-to-death-says-hamilton-psw-about-severe-shortages-of-long-term-care/article_2e62eb45-6600-5188-a9d9-de3ab67c7242.html.

Garland-Thomson, R. (2011). Misfits: A feminist materialist disability concept. *Hypatia*, 26(3), 591–609. https://doi.org/10.1111/j.1527-2001.2011.01206.x.

Garland-Thomson, R. (2015). Human biodiversity conservation: A consensual ethical principle. *The American Bioethics Journal*, 15(6), 13–15. https://doi.org/10.1080/15265161.2015.1028663.

Ghai, A. (2013). *Rethinking disability in India*. Routledge.

Gibson, E.B., & Mykitiuk, R. (2012). Health care access and support for disabled women in Canada falling short of the UN Convention on the Rights of Persons with Disabilities: A qualitative study. *Women's Health Issues*, 22(1), 111–18. https://doi.org/10.1016/j.whi.2011.07.011.

Gibson, K. (2006). The dualism of good and evil and East Indian insecurity in Guyana." *Journal of Black Studies, 36*(3), 362–81. https://doi.org/10.1177/0021934704273908.

Giubilini, A., & Sanyal, S. (2016). Challenging human enhancement. In S. Clarke, J. Savulescu, C.A.J. Coady, A. Giubilini, & S. Sanyal (Eds.), *The ethics of human enhancement: Understanding the debate* (pp. 1–25). Oxford University Press.

Glasbeek, H. (2020, June 11). Open the economy? The pandemic, costs, benefits, and capitalism. *Socialist Project: The Bullet.* https://socialistproject.ca/2020/06/open-the-economy-pandemic-costs-benefits-capitalism/.

Gleeson, B.J. (1996). A geography for disabled people? *Transactions: Institute of British Geographers, 21*(2), 387–96. https://doi.org/10.2307/622488.

Gleeson, B. (1999). *Geographies of disability.* Routledge.

Golledge, R.G. (1996). A response to Gleeson and Imrie. *Transactions: Institute of British Geographers, 21*(4), 404–11. https://doi.org/10.2307/622490.

Goodley, D. (2011). *Critical disability studies: Expanding debates and interdisciplinary engagements.* Springer.

Goodley, D. (2014). *Dis/ability studies: Theorizing disablism and ableism.* Routledge.

Government of Canada, Canadian Heritage. (n.d.) Court challenges program. Last updated 18 July 2024.https://www.canada.ca/en/canadian-heritage/services/funding/court-challenges-program.html.

Grand'Maison, V., & Lafuent E.M. (2022). Dys-Feminicide: Conceptualizing the feminicides of women and girls with disabilities. *Sociation Today, 21*(1), 129–46. https://sociation.ncsociologyassoc.org/wp-content/uploads/2022/03/dysfeminicide_final_maison-1.pdf.

Grech, S., & Soldatic, K. (Eds.). (2016). *Disability in the Global South: The critical handbook.* Springer.

Greenberg, O. (2013). The global organ trade: A case in point. *Cambridge Quarterly of Health Care Ethics 22*(3), 1–8. https://doi.org/10.1017/S0963180113000042.

Guyana improves significantly on Corruption Index – Transparency International. (2020, January 24). *Kaieteur News.* https://www.kaieteurnewsonline.com/2020/01/24/guyana-improves-significantly-on-corruption-index-transparency-intl/.

Gyngell, C., & Selgelid, M.J. (2016). Human enhancement: Conceptual clarity and moral significance. In S. Clarke, C. Savulescu, C.A.J. Coady, A. Giubilini, & S. Sanyal (Eds.) *The ethics of human enhancement: Understanding the debate* (pp. 111–26). Oxford University Press.

Hall, C.M. (2011). Health and medical tourism: A kill or cure for global health? *Tourism Review, 66*(1/2), 4–15. https://doi.org/10.1108/16605371111127198.

Hall, E. (2019). A critical geography of disability hate crime. *Area, 51*(2), 249–56. https://doi.org/10.1111/area.12455.

Hall, E., & Wilton, R. (2017). Towards a relational geography of disability. *Progress in Human Geography, 41*(6), 727–44. https://doi.org/10.1177/0309132516659705.

Hallam, K.T., Oliver, J.S., Chambers, V., & Norman, T.R. (2006). The
heritability of elatonin secretion and sensitivity to bright nocturnal light in
twins. *Psychoneuroendocrinology, 31*(7), 867–75. https://doi.org/10.1016
/j.psyneuen.2006.04.004.

Hamm, K., Schochet, L., & Nova, C. (2018). *The Trump plan to cut benefit
programs threatens children.* Center for American Progress. https://www
.americanprogress.org/wp-content/uploads/sites/2/2018/04/Hamm
YoungChildrenWillBeHarmedByBenefitsCuts-report.pdf.

Hamraie, A. (2017). *Building access: Universal design and the politics of disability.*
University of Minnesota Press.

Hansen, N., & Philo, C. (2007). The normality of doing things differently:
Bodies, spaces and geographies of disability. *Tidschrift voor Economische
Geografie, 98*(4), 493–506. https://doi.org/10.1111/j.1467-9663.2007.00417.x.

Haraway, D. (1991). A cyborg manifesto. In *Simians, cyborgs, and women: The
reinvention of nature* (pp. 149–81). Routledge.

Harpur, P. (2012). Embracing the new disability rights law: The importance
of the Convention on the Rights of Persons with Disabilities. *Disability &
Society, 27*(1), 1–14. https://doi.org/10.1080/09687599.2012.631794.

Harvey, D. (2008). The right to the city. *New Left Review, 53.* https://newleftreview
.org/issues/ii53/articles/david-harvey-the-right-to-the-city.

Harvey, D. (2013). *Seventeen contradictions and the end of capitalism.* Profile Books.

Harvey, D. (Host). (2018). *The anti-capitalist chronicles: The contradictions of
neoliberalism* [Audio Podcast]. https://www.democracyatwork.info
/acc_contradictions_of_neo_liberalism.

Hayes, M. (2018, December 10). Black people more likely to be injured and
killed by Toronto police report finds. *Globe and Mail.* https://www
.theglobeandmail.com/canada/toronto/article-report-reveals-racial
-disparities-in-toronto-polices-use-of-force/.

Healey, J. (2020). "It spreads like a creeping disease": Experiences of disability
hate crimes in austerity Britain. *Disability and Society, 35*(2), 176–200.
https://doi.org/10.1080/09687599.2019.1624151.

Hennessy, T., & Tranjan, R. (2018). *No safe harbour: Precarious work and economic
insecurity among skilled professionals in Canada.* Canadian Centre for Policy
Alternatives. https://policyalternatives.ca/sites/default/files/uploads
/publications/National%20Office%2C%20Ontario%20Office/2018/08
/No%20Safe%20Harbour.pdf.

Hidalgo, J. (2018). Cosmopolitan moral enhancement. In J. Flanigan & T.L.
Price (Eds.), *The ethics of ability and enhancement* (pp. 173–95). Palgrave
MacMillan.

Hira-Friesan, P. (2018). Immigrants and precarious work in Canada: Trends,
2006–2012. *Journal of International Migration and Integration, 19*(1), 35–57.
https://doi.org/10.1007/s12134-017-0518-0.

Hornbacher, M. (2008). *Madness: A bipolar life*. Mariner Books.

Horowitz, J.M., Igielnik, R., & Kochhar, R. (2020, January 9). Most Americans say there is too much economic inequality in the U.S., but fewer than half call it a top priority. Pew Research Center. https://www.pewresearch.org /social-trends/2020/01/09/trends-in-income-and-wealth-inequality/.

Horowitz, M.D., Rosensweig, J.A., & Jones, C.A. (2007). Medical tourism: Globalization of the health care marketplace. *MedGenMed, 9*(4), 33. https:// www.ncbi.nlm.nih.gov/pmc/articles/PMC2234298/.

Horton, H. (2023, March 9). Fossil fuels received £20bn more UK support than renewables since 2015. *The Guardian*. https://www.theguardian.com /environment/2023/mar/09/fossil-fuels-more-support-uk-than-renewables -since-2015.

Hudson, K.A. (2008). Globalization and the black market organ trade: When even a kidney can't pay the bills. *Disability Studies Quarterly, 28*(4). https:// doi.org/10.18061/dsq.v28i4.143.

Human enhancement: Is it good for society? (2019, February 11). *Science Daily*. https://www.sciencedaily.com/releases/2019/02/190211114300.htm.

Imrie, R. (1996). Ableist geographies, disablist spaces: Toward a reconstruction of Golledge's "Geography of the Disabled." *Transactions of the Institute of British Geographers, 21*(2), 397–403. https://doi.org/10.2307/622489.

Imrie, R. (2010). Disability, embodiment and the meaning of home. In V. Chouinard, E. Hall, & R. Wilton (Eds.), *Towards enabling geographies: "Disabled" bodies and minds in society and space* (pp. 23–44). Ashgate.

Inder, M.L., Crowe, M.T., Joyce, P.R., Moor, S., Carter, J.D., & Luty, S.E. (2010). "I really don't know if it is still there": Ambivalent acceptance of a diagnosis of bipolar disorder. *Psychiatric Quarterly, 81*(2), 157–65. https://doi.org /10.1007/s11126-010-9125-3.

Inder, M.L., Moor, S., Luty, S.E., Carter, J.D., & Joyce, P.R. (2008). "I actually don't know who I am": The impact of bipolar disorder on the development of the self. *Psychiatry: Interpersonal and Biological Processes, 71*(2), 123–33. https://doi.org/10.1521/psyc.2008.71.2.123.

Inequality.org. (n.d.). Global inequality. Accessed 18 August 2020, from https://inequality.org/facts/global-inequality/.

International Labour Organization. (n.d.-a). The enormous burden of dangerous working conditions. Accessed 23 July 2020, from https://www.ilo.org /moscow/areas-of-work/occupational-safety-and-health/WCMS_249278 /lang--en/index.htm#:~:text=The%20enormous%20burden%20of%20 poor%20working%20conditions&text=Worldwide%2C%20there%20are%20 around%20340,of%20work%2Drelated%20illnesses%20annually.

International Labour Organization. (n.d.-b). The enormous burden of poor working conditions. Accessed 25 January 2024, from https://www.ilo.org /moscow/areas-of-work/occupational-safety-and-health/WCMS_249278

/lang--en/index.htm#:~:text=The%20enormous%20burden%20of%20
poor,6000%20deaths%20every%20single%20day.

Jamison, K.R. (1994). *Touched with fire: Manic-depressive illness and the artistic temperament*. Simon and Schuster.

Johnson, K. (2020, June 26). Mistreatment of migrant farmworkers a "national disgrace": Health minister. *Global News*. https://globalnews.ca/news/7113934/canada-migrant-farm-workers-covid-19/.

Jones, A. (2019, December 9). Ontario government cancels future legal aid funding cuts but 2019 reductions remain. *Global News*. https://globalnews.ca/news/6273787/legal-aid-ontario-cuts/.

Kanter, A.S. (2019). Let's try again: Why the U.S. should ratify the U.N. Convention on the Rights of People with Disabilities. *Touro Law Review*, *35*(1), 1–44. https://digitalcommons.tourolaw.edu/lawreview/vol35/iss1/12.

Kayess, R., & French, P. (2008). Out of the darkness and into the light? Introducing the convention on the rights of people with disabilities. *Human Rights Law Review*, *8*(1), 1–34. https://doi.org/10.1093/hrlr/ngm044.

Klein, N. (2008). *The shock doctrine: The rise of disaster capitalism*. Vintage Canada.

Klein, N. (2015). *This changes everything: Capitalism versus the climate*. Vintage Canada.

Klein, N. (2019). *On fire: The burning case for a green new deal*. Alfred A. Knopf Canada.

Knowles, C. (2001). *Bedlam on the streets*. Routledge.

Kruse, R.J. (2010). Placing little people: Dwarfism and geographies of everyday life. In V. Chouinard, E. Hall, & R. Wilton (Eds.), *Towards enabling geographies: "Disabled" bodies and minds in society and space* (pp. 183–98). Ashgate.

La Monica, N. (2016). *Surviving or thriving in academia: Autoethnographic accounts of non-visibly disabled grads' experiences of inclusion and exclusion* [Unpublished doctoral dissertation]. McMaster University.

Lattanzio, R. (2019). Severe cuts to Legal Aid impact persons with disabilities. *ARCH Alert*, *20*(2). https://archdisabilitylaw.ca/arch_alert/arch-alert-volume-20-issue-2/#cuts-to-legal-aid.

Longhurst, R. (2010). The disabling affects of fat: The emotional and material geographies of some women who live in Hamilton, New Zealand. In V. Chouinard, E. Hall, & R. Wilton (Eds.), *Towards enabling geographies: "Disabled" bodies and minds in society and space* (pp. 199–216). Ashgate.

Lord, J.E., Suozzia, T., & Taylor, A.L. (2010). Convention on the Rights of Persons with Disabilities: Addressing the deficit in global health governance. *Journal of Law, Medicine and Ethics*, *38*(3), 564–79. https://doi.org/10.1111/j.1748-720X.2010.00512.x.

Lund, E.M., & Ayers, K.B. (2020). Raising awareness of disabled lives and health care rationing during the COVID-19 pandemic. *Psychological Trauma: Theory, Research, Practice, and Policy*, 12(S1), S210–S211. https://doi.org/10.1037/tra0000673.

Lynk, M. (2008). *The duty to accommodate in canadian workplaces: Leading principles and recent cases*. Ontario Federation of Labour. https://ofl.ca/wp-content/uploads/2008.06.21-Report-DutytoAccommodate.pdf.

Mariant, D. (2006). *Surviving bipolar's fatal grip: The journey to hell and back*. Mariant Enterprises Inc.

Martin, E. (2007). *Bipolar expeditions: Mania and depression in American culture*. Princeton University Press.

Marquis, E., Fudge Shormans, A., Jung, B., Vietinghoff, C., Wilton, R., & Baptiste, S. (2016). Charting the landscape of accessible education for post-secondary students with disability. *Canadian Journal of Disability Studies*, 5(2), 31–71. https://doi.org/10.15353/cjds.v5i2.272.

Marx, K. (2013). *Capital: Volumes 1 and 2* (S. Moore & E. Aveling, Trans.). Wordsworth Editions. (Original work published 1867)

Marx, K. (2022). *Critique of the Gotha programme* (Kevin B. Anderson & Karel Ludenhoff, Trans.). PM Press. (Original work published 1875)

May, B. (2019). *Precarious work: Understanding the changing nature of work in Canada*. House of Commons, Canada, Federal Standing Committee on Human Resources, Social Development and the Status of Persons with Disabilities. https://www.ourcommons.ca/Content/Committee/421/HUMA/Reports/RP10553151/humarp19/humarp19-e.pdf.

Maybin, S. (2019, May 8). Will Guyana soon be the richest country in the world? *BBC World Service*. https://www.bbc.com/news/world-latin-america-48185246.

McCabe, L.L., & McCabe, E.R.B. (2011). Down's syndrome: Coercion and eugenics. *Genetics in Medicine*, 13(8), 708–10. https://doi.org/10.1097/GIM.0b013e318216db64.

Mckay, S., Jefferies, S., Paraksevopoulou, A., & Keles, J. (2012). *Study on precarious work and social rights*. Working Rights Institute, Metropolitan University of London.

McRuer, R. (2018). *Crip times: Disability, globalization, and resistance*. University of New York Press.

Meekosha, H. (2011). Decolonising disability: Thinking and acting globally. *Disability and Society*, 26(6), 667–82. https://doi.org/10.1080/09687599.2011.602860.

Meekosha, H., & Soldatic, K. (2011). Human rights and the Global South: The case of disability. *Third World Quarterly*, 32(8), 1383–97. https://doi.org/10.1080/01436597.2011.614800.

Melcher, R. (2008). *Discerning bipolar grace: A man at his best dealing with bipolar at its worst*. iUniverse.

Meller, A., & Ahmed, H. (2019, August 30). *How big pharma reaps profits while hurting everyday Americans*. Centre for American Progress. https://www.americanprogress.org/article/big-pharma-reaps-profits-hurting-everyday-americans/.

Metzel, D. (2010). 563 miles: A matter of distance in long-distance caring by wiblings of siblings with intellectual and developmental disabilities. In V. Chouinard, E. Hall, & R. Wilton (Eds.), *Towards enabling geographies: "Disabled" bodies and minds in society and space* (pp. 123–43). Ashgate.

Michalak, E.E., Yatham, L.N., Kolesar, S., & Lam, R.W. (2006). Bipolar disorder and quality of life: A patient-centered perspective. *Quality of Life Research, 15*, 25–37. https://doi.org/10.1007/s11136-005-0376-7.

Mikulic, M. (2024, January 10). *The global pharmaceutical industry – statistics & facts*. Statista. https://www.statista.com/topics/1764/global-pharmaceutical-industry/#topicOverview.

Mitchell, D.T., with Snyder, S.L. (2015). *The biopolitics of disability: Neoliberalism, ablenationalism and peripheral embodiment*. University of Michigan Press.

Mladenov, T. (2015). Neoliberalism, postsocialism, disability. *Disability & Society, 30*(3), 445–59. https://doi.org/10.1080/09687599.2015.1021758.

Moniruzzaman, M. (2012). "Living cadavers" in Bangladesh. *Medical Anthropology Quarterly, 26*(1), 69–89. https://doi.org/10.1111/j.1548-1387.2011.01197.x.

Moreau, G. (2020, February 26). *Police-reported hate crimes in Canada, 2018*. Statistics Canada. https://www150.statcan.gc.ca/n1/pub/85-002-x/2020001/article/00003-eng.htm.

Morris, J. (1991). *Pride against prejudice: Transforming attitudes toward disability*. The Women's Press.

NCD calls for Disability Act to be implemented. (2018, October 28). *Guyana Times*. https://guyanatimesgy.com/ncd-calls-for-disability-act-to-be-implemented/.

Neudel, E. (Director). (2011). *Lives worth living: The great fight for disability rights* [Documentary film]. Storyline Motion Pictures.

Nordhaus, W. (2013). *The climate casino: Risk, uncertainty and economics for a warming world*. Yale University Press.

Oedegaard, K.J., Neckelmann, D., Benazzi, F., Syrstab, V.E.G., & Fasmer, O.B. (2008). Dissociative experiences differentiate bipolar II from unipolar depressed patients: the mediating role of cyclothymia and the type A behaviour speed and impatience subscale. *Journal of Affective Disorders, 108*(3), 207–16. https://doi.org/10.1016/j.jad.2007.10.018.

Oliver, M. (1990). *The politics of disablement*. Macmillan Press.

Oliver, M. (1996). *Understanding disability: From theory to practice*. Macmillan Press.

Ontario Human Rights Commission. (2021). *Human rights under pressure: From policing to pandemics: Annual Report 2020–2021*. https://www3.ohrc.on.ca /en/human-rights-under-pressure-annual-report-2020-21.

Organization of American States. (1999). Inter-American Convention on the Elimination of All Forms of Discrimination against Persons with Disabilities. https://www.oas.org/juridico/english/treaties/a-65.html.

Oswal, S.K. (2018). Disability policies and exclusionary infrastructure: A critique of the AAUP report. In D. Bolt & C. Penheth (Eds.), *Disability, avoidance and the academy: challenging resistance* (pp. 21–32). Routledge.

Pan American Health Organization and World Health Organization. (2020). *Guyana: Country Cooperation Strategy 2016-2020: Strengthening health systems to achieve universal health*. https://www.paho.org/en/documents/guyana -country-cooperation-strategy-2016-2020.

Park, D., Radford, J., & Vickers, M. (1998). Disability studies in human geography. *Progress in Human Geography, 22*(2), 208–33. https://doi.org /10.1191/030913298672928786.

Parr, H. (1999). Delusional geographies: The experiential worlds of people during madness/illness. *Environment and Planning D: Society and Space, 17*(6), 673–90. https://doi.org/10.1068/d170673.

Parr, H. (2008). *Mental health and social space: Towards inclusionary geographies*. Blackwell Publishing.

Parr, H., & Philo, C. (1995). Mapping "mad" identities. In S. Pile & N. Thrift (Eds.), *Mapping the subject* (pp. 182–207). Routledge.

Parr, H., Philo, C., & Burns, N. (2004). Social geographies of rural mental health: Experiencing inclusions and exclusions. *Transactions of the Institute of British Geographers, 29*(4), 401–19. https://doi.org/10.1111/j.0020 -2754.2004.00138.x.

Phan, S. (2020, January 22). Wealth gap widening for more than 70% of global population, researchers find. *The Guardian*. https://www.theguardian.com /global-development/2020/jan/22/wealth-gap-widening-for-more -than-70-per-cent-of-global-population-researchers-find#:~:text=Social%20 and%20economic%20disparities%20have,of%20people%20could%20be%20 damaged.

Philo, C., & Wolch, J.R. (2001). The "three waves" of research in mental health geography: A review and critical commentary. *Epidemiologia et Psiiatria Sociale, 10*(4), 130–44. https://doi.org/10.1017/S1121189X00005406.

Piepzna-Samaranmha, L.L. (2022). *The future is disabled: Prophecies, love notes and mourning songs*. Arsenal Pulp Press.

Pratt, G. (1998). Inscribing domestic work on Filipina bodies. In J.P. Jones & H. Nast (Eds.), *Places through the body* (pp. 283–304). Routledge.

Pratt, G. (2009). Circulating sadness: Witnessing Filipina mothers' stories of family separation. *Gender, Place and Culture, 16*(1), 3–22. https://doi.org/10.1080/09663690802574753.

Price, M. (2011). *Mad at school: Rhetorics of mental disability and academic life.* University of Michigan Press.

Pring, J. (2023, May 11). Errol Graham: DWP criticised by report into disabled man's starvation death. *Disability News Service.* https://www.disabilitynewsservice.com/errol-graham-dwp-criticised-by-report-into-disabled-mans-starvation-death/.

Proudfoot, J.G., Parker, G.B, Benoit, M., Manicavasagar, V., Smith, M., & Gayed, A. (2009). What happens after Diagnosis? Understanding the experiences of patients with newly-diagnosed bipolar disorder. *Health Expectations, 12,* 120–9. https://doi.org/10.1111/j.1369-7625.2009.00541.x.

Puar, J.K. (2009). Prognosis time: Towards a geopolitics of affect, debility and capacity. *Women & Performance: A Journal of Feminist Theory, 19*(2), 161–72. https://doi.org/10.1080/07407700903034147.

Puar, J.K. (2017). *The right to maim: Debility, capacity, disability.* Duke University Press.

Quinlan, M., & Bohle, P. (2009). Overstretched and unreciprocated commitment: Reviewing research on the health and safety effects of downsizing and job insecurity. *International Journal of Social Determinants of Health and Health Services, 39*(1), 1–44. https://doi.org/10.2190/HS.39.1.a.

Raso, K. (2018, July 1). Disability and the job churn. *The Monitor.* Canadian Centre for Policy Alternatives. https://policyalternatives.ca/publications/monitor/disability-and-job-churn.

Reavey P., Poole, J., Corrigall, R., Zundel, T., Byford, S., Sarhane, M., & Ougrin, D. (2017). The ward as emotional ecology: adolescent experiences of managing mental health and distress in psychiatric inpatient settings. *Health and Place, 46,* 210–18. https://doi.org/10.1016/j.healthplace.2017.05.008.

Ridge, D., Emslie C., & White. A. (2011). Understanding how men experience, express and cope with mental distress: where next? *Sociology of Health & Illness, 33*(1), 145–59. https://doi.org/10.1111/j.1467-9566.2010.01266.x.

Rihmer, Z., & Kiss, K. (2002). Bipolar disorders and suicidal behaviour. *Bipolar Disorders, 4*(Suppl. 1), 21–5. https://doi.org/10.1034/j.1399-5618.4.s1.3.x.

Rioux, M.H. (2009). Bending towards justice. In T. Titchkosky & R. Michalko (Eds.), *Rethinking normalcy: A disability studies reader* (pp. 201–16). Canadian Scholars' Press.

Robson, E. (2004). Hidden child workers: Young carers in Zimbabwe. *Antipode, 36*(2), 227–48. https://doi.org/10.1111/j.1467-8330.2004.00404.x.

Roszalski, M., Katsiyannis, A., Ryan, J., Collins, T., & Stewart, A. (2010). Americans with Disability Act Amendments of 2008. *Journal of Disability Policy Studies, 21*(1), 22–8. https://doi.org/10.1177/1044207309357561.

Russell, M. (2002). What disability civil rights cannot do: Employment and political economy. *Disability and Society, 17*(2), 117–35. https://doi.org/10.1080/09687590120122288.

Ryan, F. (2018, April 5). A year of disability cuts has done nothing but starve the sick. *The Guardian.* https://www.theguardian.com/commentisfree/2018/apr/05/year-disability-cuts-starving-sick-employment.

Salzman, L. (2010). Rethinking guardianship (again): Substituted decision making as a violation of the integration mandate of Title II of the Americans with Disabilities Act. *Colorado Law Review, 81*(1), Article 4. https://scholar.law.colorado.edu/lawreview/vol81/iss1/4.

Sanders, B., & Omar, I. (2019a, July 16). Letter to Loren Sweatt, Deputy Assistant Secretary of State Occupational Health and Safety. https://www.sanders.senate.gov/wp-content/uploads/osha-letter-amazon.pdf.

Sanders, B., & Omar, I. (2019b, July 24). Amazon warehouses: Unsafe for workers or communities. *Socialist Project: The Bullet.* https://socialistproject.ca/2019/07/amazon-warehouses-unsafe/.

Schalk, S. (2022). *Black disability politics.* Duke University Press.

Scheper-Hughes, N. (2003). Keeping an eye on the global traffic in human organs. *The Lancet, 361*(9369), 1645–8. https://doi.org/10.1016/s0140-6736(03)13305-3.

Scheper-Hughes, N. (2005). The ultimate commodity [Review of *Stakes and kidneys*, by James Stacey Taylor]. *The Lancet, 366*(9494), 1349–50. https://doi.org/10.1016/S0140-6736(05)67550-2.

Scheper-Hughes, N. (2009, October 21). Organs without borders. *Foreign Policy.* https://foreignpolicy.com/2009/10/21/organs-without-borders/.

Shakespeare, T. (2006). *Disability rights and wrongs revisited* (2nd ed.). Routledge.

Shakespeare, T. (2011). Choices, reasons and feelings: Prenatal diagnosis as disability dilemma. *Alter, 5*(1), 37–43. https://doi.org/10.1016/j.alter.2010.11.001.

Sharif, A., Singh, M.F., & Lavee, J. (2014). Organ procurement from executed prisoners in China. *American Journal of Transplantation, 14*(10). https://doi.org/10.1111/ajt.12871.

Shaver, L. (2010). Inter-American Human Rights System: An effective tool for regional rights protection? *Washington University Global Studies Law Review, 9*(4), 639–76. https://openscholarship.wustl.edu/law_globalstudies/vol9/iss4/4.

Sherry, M. (2016). *Disability hate crimes: Does anyone really hate disabled people?* Routledge.

Smith, M., & Davidson, J. (2006). "It makes my skin crawly": The embodiment of disgust in phobias of "nature." *Body and Society, 12*(1), 43–67. https://doi.org/10.1177/1357034X06061195.

Snyder, S., & Mitchell, D. (2006). Afterword–regulated bodies: Disability studies and the controlling professions. In D.M. Turner & K. Stagg (Eds.), *Social histories of disability and deformity* (pp. 175–89). Routledge.

Soldatic, K. (2013). The transnational sphere of justice: Disability praxis and the politics of impairment. *Disability & Society, 28*(6), 744–55. https://doi.org/10.1080/09687599.2013.802218.

Soldatic, K., & Grech, S. (2014). Transnationalising disability studies: Rights, justice and impairment. *Disability Studies Quarterly, 34*(2). https://doi.org/10.18061/dsq.v34i2.4249.

Southern, R. (2023, March 23). No help for struggling Ontarians, nothing for COVID in record-spending budget. *City News*. https://toronto.citynews.ca/2023/03/23/ontario-2023-budget-ford-province-spending/.

Statista. (n.d.). Number of memberships at fitness centers / health clubs in the United States from 2000 to 2022. Accessed 12 September 2023, from https://www.statista.com/statistics/236123/us-fitness-center-health-club-memberships/.

Statista Research Department. (2024). National debt of Guyana from 2007 to 2029. https://www.statista.com/statistics/1392362/national-debt-guyana/.

Statistics Canada. (2024). Police-reported hate crime, by type of motivation, selected regions and Canada (selected police services). Table: 35-10-0066-01. Last modified 18 September 2024. https://www150.statcan.gc.ca/t1/tbl1/en/tv.action?pid=3510006601.

Stienstra, D. (2011). *About Canada: Disability rights*. Fernwood Press.

Stienstra, D. (2017). DisAbling women and girls in austere times. *Atlantis: Critical Studies in Gender, Culture, and Social Justice, 38*(1), 154–67. https://journals.msvu.ca/index.php/atlantis/article/view/5330.

Stringham, A. (2007). *A Promise of hope: The astonishing true story of a woman afflicted with bipolar disorder and the miraculous treatment that cured her*. Harper Collins.

Summers, H. (2020, June 30). COVID-19 intensifies elder abuse globally as hospitals prioritize young: Older patients turned away or left untreated, while domestic abuse is also rising. *The Guardian*. https://www.theguardian.com/global-development/2020/jun/30/covid-19-intensifies-elder-abuse-globally-as-hospitals-prioritise-young.

Swann, A.C., Moeller, F.G., Steinberg, A.L., Schneider, L., Barratt, E.S., & Dougherty, D.M. (2007). Manic symptoms and impulsivity during bipolar depressive episodes. *Bipolar Disorders, 9*(3), 206–12. https://doi.org/10.1111/j.1399-5618.2007.00357.x.

Syed, F. (2019, July 16). Here's everything the Doug Ford government cut in its first year in office. *National Observer*. https://www.nationalobserver.com/2019/06/07/news/heres-everything-doug-ford-government-cut-its-first-year-office.

Thomas, C. (1999). *Female forms: Experiencing and understanding disability*. McGraw-Hill Education.

Thompson, A. (2019, December 20). There is something fundamentally wrong with the Guyana Police Force. *Stabroek News*. https://www.stabroeknews

.com/2019/12/20/features/the-minority-report/there-is-something
-fundamentally-wrong-with-the-guyana-police-force/.

Titchkosky, T. (2008). "To pee or not to pee?": Ordinary talk about extraordinary exclusions in a university environment. *Canadian Journal of Sociology, 33*(1), 38–60. https://doi.org/10.29173/cjs1526.

Titchkosky, T. (2011). *The question of access: Disability, space, meaning*. University of Toronto Press.

Titchkosky, T., & Michalko, R. (2009). Introduction. In T. Titchkosky & R. Michalko (Eds.), *Rethinking normality: A disability studies reader* (pp. 1–14). Canadian Scholars' Press.

Tucker, I. (2010). The potentiality of bodies. *Theory & Psychology, 20*(4), 511–27. https://doi.org/10.1177/0959354309360947.

Tunnicliffe, J. (2006). *The Ontario Human Rights Code review, 1975–1981: A new understanding of human rights and its meaning for public policy* [Unpublished master's thesis]. University of Waterloo.

Tyner, J.A. (2016). *Violence in capitalism: Devaluing life in an age of responsibility*. University of Minnesota Press.

United Nations Development Programme. (2022). *Human development report 2021–22: Uncertain times, unsettled lives: Shaping our future in a transforming world*. https://hdr.undp.org/content/human-development-report-2021-22.

United Nations Global Issues. (n.d.). Climate change. Accessed 10 December 2019, from https://www.un.org/en/global-issues/climate-change.

United Nations Human Rights. (n.d.). Guyana – IMM Situation. Committee on the Rights of Persons with Disability. Accessed 17 September 2023, from https://www.ohchr.org/en/treaty-bodies/crpd/guyana-imm-situation.

US Bureau of Labor Statistics. (2019, March 1). Employment of people with a disability in 2018. *The Economics Daily*. https://www.bls.gov/opub/ted/2019/employment-of-people-with-a-disability-in-2018.htm.

US Bureau of Labor Statistics. (2023). Table A-6. Employment status of the civilian population by sex, age, and disability status, not seasonally adjusted [Economic News Release]. https://www.bls.gov/news.release/empsit.t06.htm/.

US Citizenship and Immigration Services. (2012). *Overview of INS history*. USCIS History Office and Library. https://www.uscis.gov/sites/default/files/document/fact-sheets/INSHistory.pdf.

US Department of Justice. (n.d.). Americans with Disabilities Act 1990, As Amended. Accessed 25 January 2024, from https://www.ada.gov/law-and-regs/ada/.

US Government Accountability Office. (2022, March 2). Science & tech spotlight: Long COVID. https://www.gao.gov/products/gao-22-105666.

Valentine, G. (2007). Theorizing and researching intersectionality: A challenge for feminist geography. *The Professional Geographer, 59*(1), 10–21. https://doi.org/10.1111/j.1467-9272.2007.00587.x.

Valero, M.V. (2023). Death threats, trolling and sexist abuse: Climate scientists report online attacks. *Nature, 616*(7957), 421–2. https://doi.org/10.1038/d41586-023-01018-9.

Walker, D., & Florence, M. (2006). Grounded theory: An exploration of process and procedure. *Qualitative Health Research, 16*(4), 547–59. https://doi.org/10.1177/1049732305285972.

Weiland, M.F., with Larkin, W. (2009). *Fall to pieces: A memoir of drugs, rock 'n' roll, and mental illness*. William Morrow.

Wendell, S. (1996). *The Rejected body: Feminist philosophical reflections on disability*. Routledge.

Wilton, R., & Evans, J. (2016). "Social enterprises as spaces of encounter for mental health consumers." *Area, 48*(2), 236–43. https://doi.org/10.1111/area.12259.

Wilton, R., DeVerteuil, G., & Evans, J. (2013). "No more of this macho bullshit": Drug treatment, place and the reworking of masculine identity. *Transactions of the Institute of British Geographers, 39*(2), 291–303. https://doi.org/10.1111/tran.12023.

Wilton, R., Hansen, S., & Hall, E. (2017). Disabled people, medical inadmissibility, and the differential politics of immigration. *Canadian Geographies, 61*(3), 389–400. https://doi.org/10.1111/cag.12361.

Wolbring, G. (2008). The politics of ableism. *Development, 51*, 252–8. https://doi.org/10.1057/dev.2008.17.

World Bank Group. (2020). *Systematic country diagnostic: A pivotal moment for Guyana – realizing the opportunities*. https://www.worldbank.org/en/region/lac/publication/world-bank-group-systematic-country-diagnostic-a-pivotal-moment-for-guyana-realizing-the-opportunities.

World Health Organization. (2020). *Report on cancer: Setting priorities, investing wisely and providing care for all*. World Health Organization.

World Wildlife Fund. (n.d.) Polar bear: Facts. Accessed 18 August 2020, from https://www.worldwildlife.org/species/polar-bear.

Worth, N., Simard-Gagnon, L., & Chouinard, V. (2017). Disabling cities. In A. Bain & L. Peake (Eds.), *Urbanization in global context* (pp. 1–33). Oxford University Press.

Young, I.M. (1990). *Justice and the politics of difference*. Princeton University Press.

Zannolli, L. (2019, December 11). "We're just waiting to die": The Black residents living on top of a toxic landfill site. *The Guardian*. https://www.theguardian.com/us-news/2019/dec/11/gordon-plaza-louisiana-toxic-landfill-site.

Zucman, G. (2019). Global wealth inequalities. *Annual Review of Economics, 11*, 109–38. https://doi.org/10.1146/annurev-economics-080218-025852.

Index

www.ingramcontent.com/pod-product-compliance
Lightning Source LLC
Chambersburg PA
CBHW031124020426
42333CB00012B/219